Assessment in the
College Science Classroom

Assessment in the College Science Classroom

Clarissa Dirks
The Evergreen State College

Mary Pat Wenderoth
University of Washington

Michelle Withers
West Virginia University

Part of the
W.H. Freeman Scientific Teaching Series

W. H.
FREEMAN
www.whfreeman.com

Publisher:	Susan Winslow
Editor:	Sara Ruth Blake
Development Editor:	Anna Bristow
Cover and Text Designer:	Mark Ong, Side By Side Studios
Production Coordinator:	Ellen Cash
Composition:	Susan Riley, Side By Side Studios
Printing and Binding:	RR Donnelley

Scientific Teaching Series Editor: William B. Wood, Science Education Initiative and Department of Molecular, Cellular, and Developmental Biology, University of Colorado, Boulder, CO

Scientific Teaching Series Editor: Sarah Miller, Madison Teaching and Learning Excellence, University of Wisconsin-Madison, Madison, Wisconsin

Cover photo credit: Jill Tindall

Library of Congress Control Number: 2013951629

ISBN-13: 978-1-4292-8197-3
ISBN-10: 1-4292-8197-9

Printed in the United States of America

First printing

W.H. Freeman and Company
41 Madison Avenue
New York, NY 10010
Houndmills, Basingstoke
RG21 6XS, England
www.whfreeman.com

Contents

Foreword   vii

Preface   xi

SECTION I: FRAMEWORK FOR ASSESSMENT   1

1 Using Assessment for
Effective Instructional Design   3

2 Evaluating the Cognitive Levels of Instructional
Materials Using an Educational Taxonomy   14

SECTION II: ASSESSMENT IN PRACTICE   41

3 Summative Assessment: Assessment of Learning   43

4 Formative Assessment: Assessment for Learning   60

5 Assessing Higher-Order Cognitive
Skills with Multiple-Choice Questions   74

6 Preparing Students for Assessment   92

SECTION III: ASSESSMENT BEYOND THE CLASSROOM   105

7 Assessing the Effectiveness of Teaching Innovations   107

8 Assessment Workshop: Disseminating What You Have Learned   124

Glossary of Pedagogical Terms   135

References   141

Index   159

About the Authors   169

Foreword

Jay B. Labov, PhD

In his magnum opus *The Scarlet Letter*, Nathaniel Hawthorne relates the tale of Hester Prynne and the scarlet *A* that she was required to wear. The *A* was an onus to be avoided at all costs by those who would be forced to wear it because it engendered ridicule from the community.

I envision this classic when observing what many in higher education see as modern-day Scarlet *A*s: Assessment and Accountability. Faculty often equate assessment and accountability with unwanted and unnecessary intrusions on their freedom to teach courses as they deem appropriate, and some leaders of higher education institutions may view them as an encroachment on academe's long tradition of independence, unfettered by political and other outside forces.

Academic integrity and independence are cherished concepts in higher education. They are continually challenged and must be fiercely guarded. They are central to the continued health of higher education and the services it provides to modern society. But institutions of higher education, especially public colleges and universities, have an obligation to students, the parents who support them, the public, and policymakers to demonstrate that their educational missions are being fulfilled and that their students are being well-served. These are tensions that many see as intractable, perhaps even irreconcilable.

However, as the authors of this important new book amply demonstrate and document, assessment and accountability need not be viewed as anathema to the goals of higher education. Instead, the authors show that well-designed and strategically implemented assessment can serve to address the needs of many stakeholders simultaneously. If educators and education policymakers properly understand and differentiate the goals of formative and summative assessment, students (especially undergraduates), faculty members, university administrators, and those who oversee higher education will benefit.

Research on human cognition indicates that people must practice the skills for which they will later be held accountable. Formative assessment reinforces and rewards learning goals. Ongoing formative assessment enables students to more easily check their progress, recognize gaps in their learning throughout the term, and make adjustments to their learning strategies much more effectively than is possible with high-stakes tests that are administered only once or twice in a term. They can then better modify their approaches to studying before they encounter major and often uncorrectable academic problems.

Faculty members can use formative assessment to learn about problems that individual students or the class as a whole may be experiencing. They then can intervene early to correct problems. This is especially important in science courses, in which concepts build on each other over time. In addition, clear evidence of gains in student learning is inherently rewarding to educators who too often labor to teach without really knowing whether students are benefitting. And this evidence can be especially useful when faculty members are themselves subjected to summative assessments such as tenure, promotion, or salary increases. Armed with data about student learning gains, faculty, academic departments and programs, and ultimately institutions will be able to chart their own course for improving student learning while also demonstrating to both policymakers and the public their commitment to student learning, recognition of gaps in that learning, and processes for mitigating deficiencies.

Faculty are busy and often overcommitted people. Suggesting that they incorporate assessments into their teaching beyond those traditionally given during a semester may seem ludicrous (or at best, naive). However, as this book emphasizes throughout, while an ongoing plan for assessment of student learning that is well designed and seamlessly integrated with learning goals and course design may initially require extra time and effort, the rewards (both obvious and intrinsic) are well worth the effort. With assistance from this volume, the "activation energy" required to understand, implement, and then sustain this process will become much lower.

This book is part of a series on undergraduate science education that began with *Scientific Teaching* by Jo Handelsman, Sarah Miller, and Christine Pfund. This series provides a broad vision and plan of action for both current and future faculty who wish to change how they teach, with the goal of instilling in a much broader spectrum of students a love and enthusiasm for learning science.

Given today's economic and political climates, if faculty and the institutions they serve do not understand, accept responsibility for, and then work proactively and collectively to implement a coherent plan for assessment and accountability that is based on evidence about how people learn, almost certainly such requirements will be imposed from outside of academe. The authors provide important background, justification, and methods for undertaking this process. But beyond providing needed accountability, systems of assessment that are seamlessly integrated into the fabric of higher education can also change the ways that we think about the nature of education itself.

This book provides the evidentiary base for this change in thinking, plus practical ideas for getting started. Indeed, even those faculty members who remain unconvinced about the need for greater emphasis on assessment will find practical suggestions for improving their teaching, such as improving the writing of multiple-choice questions or developing intended learning outcomes as the basis for constructing a course syllabus. My prediction is that current skeptics who read this book will find themselves thinking differently about how they teach, how their students learn, and how these two processes are inextricably connected.

Unlike traditional notions of the Scarlet *A*, the *As* of Assessment and Accountability are to be embraced. This book will enable readers to do so more readily.

Preface

The Revolution Continues: Training Two-Handed Pianists

Assessment in college science courses measures student learning. So why write an entire book on the subject? Because assessment can be used for much more than evaluation! The real potential of assessment lies in its ability to enhance learning. In addition to providing feedback on learning, assessment can be used to embed the true nature of science in the classroom. In passive lecture classes, students miss the opportunity to practice necessary critical thinking skills. By putting students, rather than instructors, into the role of question-answerer, data-interpreter, and problem-solver, assessment helps students learn content *while* practicing the scientific process and developing the critical thinking skills necessary to be a scientist.

Scientific Teaching, the inaugural book in the Scientific Teaching series, opens with an analogy likening graduate schools, which train science students in research but not teaching, to music schools that would teach students to play piano with their right hand only, presuming that the left hand will eventually "figure it out." A common assumption is that researchers who understand science should be able to teach it as well. But practically anyone who has participated in a class, whether as a teacher or a student, can attest to the fact that knowing something and being able to effectively teach it are two different things. Without formal training in different teaching methods, college science instructors often adopt the same lecture style of teaching they experienced as undergraduates. The demands of an academic position leave little time for faculty to navigate the education literature and learn the most effective teaching strategies. In an effort to give faculty a portal into these theories and practices, Jo Handelsman, Sarah Miller, and Christine Pfund wrote *Scientific Teaching* (2007).

Scientific teaching, as an approach, distills the best of modern educational theory and practice as it pertains to science education, with a unique twist. It calls on scientists to apply the process and rigor of their research to teaching, and to infuse that scientific way of thinking into the students' curriculum. When determining which experimental method will be most effective in answering a question, researchers consult the literature, discuss it with colleagues, and gather data to determine the utility of that technique. Why not use these practices to benefit the classroom? Scientific teaching encourages instructors to rely on educational literature to guide instructional choices, and challenges instructors to collect their own data—experimental, observational, or reflective—about which practices lead to success in the classroom. Classroom assessment allows instructors to collect such data using the same evidence-based practices that researchers are familiar with in the laboratory or field. More importantly, it provides feedback that moves students toward scientific understanding. The beauty of scientific teaching is that it charges the instructor simultaneously to approach teaching like a scientist and to teach students the practices that are hallmarks of that science.

This book extends the mission of *Scientific Teaching*, putting a spotlight on assessment in the college classroom. In that spirit, we have attempted to distill the best assessment methods from the literature and provide practical guidance to implement them. Our primary goal is to help college science faculty take assessment beyond the role of assigning grades and use it to its fullest potential. Throughout the book, we have integrated a variety of assessment techniques with the latest advances from cognitive research to construct effective approaches that are readily accessible to faculty. We envision science faculty with various levels of teaching experience using this book in a variety of ways. Some may wish to read the book from start to finish as a primer on the scholarship and uses of assessment in effective course design. Others may wish to target specific topics, in which case the chapters can stand alone even though they build on one another. In most chapters, we include tables with examples of assessment questions or activities from several scientific disciplines for those looking for specific templates upon which to create their own assessments.

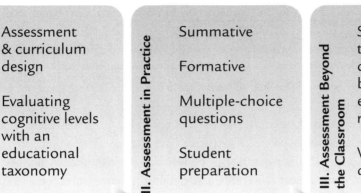

The topics in this book are organized into three sections: (I) Framework for assessment, (II) Assessment in practice, and (III) Assessment beyond the classroom. The figure above provides a visual overview of the topics in each section.

The two chapters in Section I provide a framework for using assessment to inform effective curricular design. Chapter 1 offers an introduction to the types and roles of assessment, with an emphasis on using assessment to design instructional materials that focus on intended student learning outcomes. Chapter 2 demonstrates how to use an educational taxonomy to determine the cognitive levels of instructional materials and to evaluate alignment between instruction, assessment, and desired learning outcomes.

The chapters that comprise Section II offer practical ways to evaluate and drive student learning toward the outcomes identified in Section I. The section begins with Chapter 3, which provides an overview of the familiar—"summative" assessments—and offers the reader a toolbox of methods and techniques to evaluate learning. Chapter 4 details how to use regular, ongoing assessment to foster learning and explains how to engage students in assessing their *own* learning. Chapter 5 addresses one of the most-used types of assessment, multiple-choice questions, offering tips, techniques, and advice for avoiding pitfalls. This section closes with Chapter 6, which presents strategies to help students better understand their own learning and prepare for the coming changes in assessment in the classroom.

Section III allows the reader to explore assessment beyond its uses in classroom instruction. Chapter 7 provides a primer on how to move from effective classroom assessment to curricular evaluation and discipline-based education research. Finally, in Chapter 8, we provide a sample workshop on assessment for those who wish to further advance their learning by sharing the approaches with their colleagues.

By the end of this book, we hope our readers will view assessment as one of the most powerful tools an instructor has to influence learning. We anticipate that readers will find that designing learning environments is not as daunting as it may first appear. Yes, using assessment to create an effective scientific classroom takes work, but the reward is improved student learning, which is ultimately the point.

A Note on Terminology Usage in This Book

Education and assessment vocabulary can be confusing because often terminology is used interchangeably when, in fact, there are subtle but important differences between specific words. For example, during professional development workshops it has been the authors' experience that many science faculty confuse the terms "goals," "learning objectives," and "learning outcomes," and use them interchangeably. This is no surprise because those running professional development workshops for faculty often use all three of these terms, associating them with specific definitions from the literature, and sometimes even indicate that the terms are interchangeable; this has been well-documented in other settings (Rowntree 1982; Otter 1992; Harden 2002). However, the use of these terms is so contentious that entire articles have been written to clarify the differences between the terms and their use (Allan 1996; Harden 2002). Therefore, we would like to briefly clarify why we use particular terminology to describe certain processes or behaviors.

The use of the word "objective" can be traced to descriptions in the late 1940s of how institutions, education programs, and teachers affect student learning (Mager 1962; Popham *et al.* 1969; Tyler 1969; MacDonald-Ross 1973; Cohen and Manion 1977). In these descriptions, the word "objective" in any context was used to mean the intent of what a student would learn as a result of the "input" of instruction.

In considering the multiple ways in which the word "objective" was used to describe different kinds of inputs, Eisner further separated objectives to include the category of "expressive outcomes," or what the student achieves during the learning process rather than what the instructor intended to teach (Eisner 1979). The definition was further refined to "learning outcome," meaning "what a learner knows or can do as a result of learning" (Otter 1992). The current use of "learning outcome" in higher education focuses on subject-based outcomes that are student-centered, highlighting the "output" of learning (Allan 1996).

Learning outcomes are measured through assessment. For learning purposes faculty can articulate intended learning outcomes; these often mirror learning objectives, since learning and teaching are closely connected. When faculty write learning outcomes, they are emphasizing the role of the learner. Keeping that perspective throughout the process is intended to guide instruction away from the more traditional input-based view to a more output- and student-centered view of learning (Otter 1992; Harden 2002). This change also echoes the current shift toward outcome-based education, as seen in national standards and competencies based on learning outcomes.

Many chapters of this book are dedicated to how we measure what students have learned and how we can tailor our teaching to best help them learn. From this student-centered perspective, we would like to define the following terms and how we will use them.

Learning Goals: These are general statements of what students are intended to learn as a result of the instructional experience. Goals are often stated using words or phrases that are not directly measurable, such as "understand," "know," "appreciate," and "critical thinking." Often these are the goals the instructor intends for the student, but students may also be included in the development of the goals.

Learning Objectives: In common usage, "learning objectives" can be defined as statements of what the learner should know and be able to do as a result of instruction: the intended learning outcomes. However, since historically these statements were written from the instructor's perspective and emphasized what the instructor intended the student to learn, some educators raised concerns that this term overemphasized the teaching process, or input, rather than student learning (see Allan 1996). Moreover, the term "learning objective" can be misleading because "objective" is synonymous with "goal" or "aim." To avoid this confusion, we will refrain from using this term and instead use *intended learning outcomes* as described below.

Intended Learning Outcomes: The dictionary's definition of "learning outcome" generally refers to what a learner knows and can do as a result of learning during the instructional experience. However, education researchers have advocated altering this meaning, stating that "learning outcomes subsume learning objectives" (Allan 1996). In this interpretation, outcomes refer not only to the actual results, but also the intended results of instruction. For example, it has been proposed that faculty need to shift their emphasis from *teaching* to *student learning* (Barr and Tagg 1995). Therefore, outcomes can be formulated at the start of instructional design. This redefinition is intended to help faculty shift their focus from instruction to what students actually learn.

A drawback to this use of the term is that it goes against standard English usage and no longer distinguishes desired outcomes and the actual results of assessment. To avoid this confusion, we use the term *intended learning outcomes* throughout this book. Articulating intended learning outcomes at the start of a course makes the instructional process more transparent, while giving ownership of the learning to the student. This relationship between assessment, learning, and teaching is best described by J. Allan and is the theoretical framework for backward design (Allan 1996; Wiggins and McTighe 1998).

> The more subject-specific, personal, transferable, and academic outcomes are clearly expressed, the more the learner is able to concentrate on what he/she needs to know in order to succeed on a given module or course. This places a greater emphasis on the specification of assessment tasks and the criteria by which judgments will be made, thereby forcing both the student and the teacher to examine and articulate the relationship between learning outcomes, assessment, and the experience of learning.
>
> —J. Allan (1996)

I

FRAMEWORK FOR ASSESSMENT

Our goal is to provide college science instructors with the rationale, framework, and tools to use assessment to its fullest extent in order to improve learning. The chapters in this section offer a framework for effectively embedding assessment into college course design. By necessity, the information in these chapters is not comprehensive; rather, it provides an overview of recommended practices that have emerged from the educational research about the role of assessment in the classroom. The guidelines put forward in the first two chapters serve as a foundation for the practical applications that are the subjects of the chapters in Sections II (Assessment in Practice) and III (Assessment Beyond the Classroom).

1

Using Assessment for Effective Instructional Design

If you don't know where you are going, any road will get you there.

—Lewis Carroll

Assessment is arguably the most powerful tool to improve teaching and learning that educators have at their disposal (Broadfoot and Black 2004). At its most effective, assessment is a continuous and systematic collection of data to evaluate and enhance student learning (Marchese 1993; Angelo 1995; Palomba and Banta 1999; Pellegrino *et al.* 2001; Robbins 2009). Frequent assessment provides evidence for learning in much the same way that experiments provide data for research, which should make the assessment strategies presented in this book familiar to science faculty. Throughout the book, we integrate key assessment techniques with the latest advances from human cognition research to construct a framework for best practices of assessment. In this chapter, we begin with an overview of the forms of assessment and how they support effective instructional design.

Forms of Assessment

In the college science classroom, assessment serves two critical functions: evaluating *and* facilitating student learning. These functions are fulfilled by summative and formative assessment (Table 1.1). Both summative and formative assessment have powerful impacts on learning (Wiliam and Black 1996; Black 1998). **Summative** assessment activities, such as exams or papers, usually occur at the end of a session, unit, or term to evaluate what students have learned. They are often used to determine students' grades, making them high-stakes exercises that influence student behavior and drive learning accordingly. Summative assessment can also take the form of national standardized tests such as the scholastic aptitude test (SAT) and graduate record exam (GRE), which are used to determine whether students are academically prepared to proceed to the next level of their education. Since summative assessments occur at the end of a learning period, the feedback they provide to students may prove less useful than if the assessments had occurred during the learning process, when students would have had the opportunity to revisit material or retool their study habits. We discuss in Chapter 3 how summative assessment activities can be used more effectively as learning tools.

In contrast, **formative** assessment activities occur during the learning process and offer immediate feedback to both students and instructors. These assessments include frequent testing and opportunities for recalling or retrieving information from memory. The stakes for these assessments are often low, typically worth few or no points, allowing students a safe opportunity to test their current understanding. Everyone can benefit from formative assessment activities. For example, formative assessments allow instructors to gauge and respond to the students' performance. In addition, the regular practice and feedback can also help students modify and focus their studying. As such, formative assessment helps students become aware of their own learning, begin to understand it, and thus gain some measure of control over it, a process termed "metacognition." Substantial evidence from the learning sciences indicates that helping students recognize their misconceptions, adjust their conceptual frameworks, and develop their metacognitive skills improves their learning (Bransford *et al.* 1999).

Table 1.1 General Aspects of Summative and Formative Assessment

	Summative assessment	Formative assessment
Relationship to learning	Assessment *of* learning	Assessment *for* learning
Purpose for faculty	Assign course grades Validate and accredit programs	Diagnose student learning issues Inform changes to teaching
Purpose for students	Evaluate learning Enter programs Graduate from programs Achieve professional licensing	Improve metacognition Modify study behavior
Graded	Usually	Sometimes
Timing	Periodic	Frequent
Feedback	Delayed feedback, usually as a score or grade on an assessment	Immediate feedback, usually with explanation and opportunity for reflection
Stakes	Usually high stakes One opportunity to do well	Usually low stakes Multiple opportunities to improve
Common examples	Exams Portfolios Presentations Written reports	Group problem-solving Homework In-class clicker questions Minute papers Draft versions of presentations or reports

Assessment Informs Instructional Design

Assessment plays a central role in learning, so why not let it play a central role in instructional design as well? A traditional "coverage" approach to curriculum development relies on the assumption that students will learn content by being exposed to it. In this format, instructors are the active participants and therefore the learners; they define key terms, explain fundamental concepts, and present solutions to problems, while the students, as passive recipients of the material, dutifully transcribe notes. Assessment, in turn, becomes an afterthought. Studies reveal, however, that students do not achieve the desired mastery of the subject from such a passive approach (Knight and Wood 2005).

In contrast, a "backward design" approach to instruction emphasizes what students will *learn*, not what instructors will *teach*. This approach is termed "backward" because it begins by identifying the elements that are most essential for students to know or be able to do by the end of the instructional experience: the **intended learning outcomes** (Wiggins and McTighe 1998).

The second step in backward design is **assessment**. Using the intended learning outcomes as a lodestar, instructors can design the instruments that will be used to determine whether and to what extent students are achieving—or making progress toward—those outcomes.

A clear vision of how to assess intended learning outcomes leads to the third step in backward design: **planning the instruction**. Well-designed assessment tools allow instructors to create a more effective learning environment that focuses on outcomes. In this way, assessment becomes a driving force for instructional design—and subsequently for learning—rather than an afterthought in teaching.

What does this mean for the students? Instruction designed "backward" marries intended learning outcomes with assessment to create classrooms where students understand the learning expectations and how to meet them. This demystifies for students the instructor's intentions for the learning experience, and simplifies the instructor's decisions about what to include and what to jettison—the fourth step in backward design: **evaluation of the instruction**. By providing instructional transparency, backward design naturally increases communication and reduces frustration for everyone. Figure 1.1 shows a modified version of the basic steps of backward design.

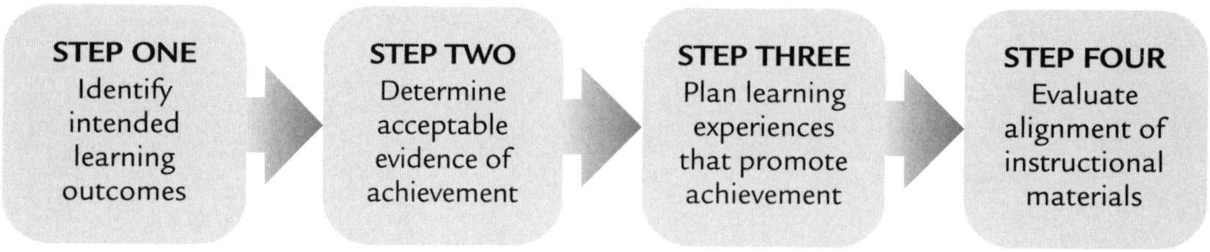

| **STEP ONE** Identify intended learning outcomes | → | **STEP TWO** Determine acceptable evidence of achievement | → | **STEP THREE** Plan learning experiences that promote achievement | → | **STEP FOUR** Evaluate alignment of instructional materials |

Figure 1.1. Backward Design (adapted)

Backward Design in Practice

The critical element of backward design is its focus on intended learning. The simplicity of this approach makes it suitable for all disciplines, class sizes, student levels, and teaching styles. When decisions about instruction and assessment are driven by intended learning outcomes, there are as many ways to help students learn as there are imaginative instructors. While the specifics of a backward-designed course can vary greatly, the following four questions are invaluable when implementing the general approach:

▶ What should students know or be able to do by the end of the course?
▶ How will the instructor know if students achieve the intended learning outcomes?
▶ What instructional approaches will maximize the likelihood that students will achieve the intended learning outcomes?
▶ Does the assessment match the instruction?

Identify Intended Learning Outcomes

During the making of *A Private Universe*, a documentary about how misconceptions impede learning, interviewers asked students and faculty what factors cause the Earth's seasons. Undergraduates, graduates, and even professors routinely gave the same wrong answer: the distance of the Earth from the sun at different times of the year (Schneps *et al.* 1989). This was the same misconception stated by high school students prior to any study of planetary motion.

Think about your own teaching: if students were being interviewed at the end of your course, what would you be embarrassed to find they didn't know? This sobering question can bring clarity to decisions about which concepts and skills constitute meaningful learning in your classroom. When intended learning outcomes address knowledge, conceptual understanding, skills, and attitudes in addition to content, they provide the most comprehensive roadmap to learning (see Appendix B for a backward design worksheet). Ask the following questions when evaluating whether or not material should be included in the intended learning outcomes for a course:

▶ Is it likely that students will retain the information, concepts, or skills long-term?

▶ Is the material necessary to understand or gain other information, concepts, or skills?

▶ What are the consequences of not retaining the information, concepts, or skills?

If the answers to these questions are "no," "no," and "nothing,"then consider replacing those intended learning outcomes with ones that reflect acquisition of crucial conceptual understanding or thinking skills.

Many factors influence decisions about intended learning outcomes for a course. If the course fulfills a degree requirement or serves as a prerequisite for other courses, it may be necessary to get input from other faculty. If the course is part of a core curriculum, the foundational courses required for all majors in a discipline, decisions about intended learning outcomes should involve guidance from that department. Instructors may also find it helpful to seek advice from external sources such as published reports from the National Research Council (2012a), the American Association for the Advancement of Science (2007; 2011), the Partnership for 21st Century Skills (2009), the Association of American Medical Colleges and the Howard Hughes Medical Institute (2009), and the College Board (2009), to name a few. In addition, several science disciplines have published inventories of questions that assess core concepts for that subject matter (see Chapter 7, Appendix A). Together, these resources can provide invaluable information for the writing of intended learning outcomes that drive meaningful learning of content and concepts, as well as science process and reasoning.

Determine Acceptable Evidence of Achievement

What evidence will demonstrate that students have achieved the intended learning outcomes? The answer to this question guides the creation of effective summative assessments (see Chapter 5 for more on designing effective exam questions). For example, assessment activities that demonstrate whether students can apply the scientific method would differ greatly from those that test whether students can define the scientific method. While both "applying" and "defining" fall under the broad goal of "understanding the scientific method," they are quite different in the specific knowledge and skills necessary to carry them out. Because of the ambiguity inherent in verbs like "know," "understand," and "appreciate," using these verbs to evaluate achievement of broad goals is difficult and subjective. In contrast, assessing specific intended learning outcomes such as "describe plate tectonics" or "predict the outcome of adding sodium to liquid water" is much more straightforward (see Appendix A for discipline-specific examples). Therefore, restating broad learning goals as more explicit intended learning outcomes leads unambiguously to the creation of metrics for assessment.

Plan Learning Experiences That Promote Achievement

Studies show that students who actively engage in the subject matter learn more than students who simply listen to lectures (Beichner and Saul 2003; Knight and Wood 2005; Freeman *et al.* 2007; Walker *et al.* 2008). Backward-designed courses in particular engage students in learning activities specifically created to help them achieve the intended learning outcomes. Learning activities that encourage students to construct their own understanding or to practice critical-thinking skills provide feedback to help gauge student progress toward the intended learning outcomes (Black and Wiliam 1998). In this way, learning activities serve as formative assessment (see Chapter 4 for further discussion of this point). Since class time is limited, instructors must decide which intended learning outcomes to address in class. Activities that pose the biggest mental challenges warrant class time, while acquisition of simpler content can occur through out-of-class assignments (see Chapter 4 for more on in- and out-of-class learning activities). For example, learning to effectively apply the scientific method requires iterative, guided practice, whereas learning to define the scientific method can easily be done on one's own.

Evaluate Alignment of Instructional Materials

Alignment between intended learning outcomes, learning activities, and assessments is critical. Intended learning outcomes and summative assessment questions can span many levels of difficulty, from simple comprehension to evaluation of concepts; therefore, learning activities should be designed to reach the same level of difficulty (Stiggins *et al.* 2007). Performance on an exam has been shown to improve when the level of thinking required during the exam matches the level of thinking required during class. Cognitive scientists refer to this phenomenon as "transfer-appropriate processing" (Morris *et al.* 1977; McDaniel *et al.* 1978; Jacoby 1983; Roediger and McDermott 1993; Thomas and McDaniel 2007). In this book, we refer to it simply as "alignment."

Because summative assessment activities generally determine a large proportion of the course grade, they have a strong influence on what students spend time studying (Morgan *et al.* 2007; Wormald *et al.* 2009). If the assessments are not aligned with class activities and outcomes, students will likely become frustrated. For example, if exams require factual recall, while intended learning outcomes and activities emphasize deep conceptual understanding, students will memorize facts and disregard the learning activities, as indicated by the common student question: "Will this be on the test?" Conversely, when students are expected to demonstrate complex skills on exams but sit passively in class, the mismatch between expectations and training leads to poor performance and frustration.

Resources like Table 1.2 provide a useful format to evaluate the alignment of outcomes and assessment. Appendix A offers examples of summative and formative assessment activities aligned to intended learning outcomes for a variety of scientific disciplines. In addition, Chapter 2 provides in-depth guidance for evaluating and aligning the cognitive levels of instructional materials using an educational taxonomy.

Table 1.2 Backward Design in Practice: How to Align Elements of Instruction

Modified from Table 5.3 in *Scientific Teaching* (Handelsman *et al.* 2007)

Step 1 Identify intended learning outcomes	Step 2 Determine acceptable evidence of achievement	Step 3 Plan learning experiences that promote achievement	Step 4 Evaluate alignment of instructional materials
Intended learning outcomes should be stated in terms that can be clearly measured by the summative assessments. For example, "students will understand the scientific method" is too broad to be easily assessed, while "students will be able to formulate a testable hypothesis" leads to an obvious assessment.	Summative assessments should provide evidence about how well students have achieved the intended learning outcomes, at a level appropriate for their developmental stage. For example, college students who achieved the intended learning outcome in Step 1 should be able to start with a set of observations and construct tentative explanations that could be supported or refuted with evidence.	Learning activities should be designed specifically to help students achieve the intended learning outcomes. For example, for students to achieve the intended learning outcome in Step 1, they should engage in learning activities that let them practice formulating and designing tests of hypotheses, with ample substantive feedback on their progress.	Each intended learning outcome and its corresponding activities and assessments should be at the same cognitive level. For example, if students simply hear a lecture defining the scientific method but are expected to perform the example in Step 2 on the exam, then the learning experience is not aligned with the assessment.

Backward Design Beyond the Classroom

After experiencing the benefits of backward design in the context of a single course, it becomes obvious that this approach works equally well at a higher level, such as designing or revising curricula for degree programs. It isn't trivial for a department to agree on what students should know and be able to do when they graduate. It is, however, a very beneficial exercise for faculty because it encourages them to identify the core concepts of their discipline. It also compels faculty to move beyond decisions about discipline-specific content and to set clear outcomes for skills and practices that students should master as part of their learning experience in the overall program. Intended learning outcomes for programs of study can be powerful tools for guiding decisions about which courses will address the specific outcomes and constitute the core curriculum. Assessments that allow departments to demonstrate the achievement of intended learning outcomes of the curriculum will be important tools for continuing program evaluation both internally and for accreditation purposes.

Conclusion

Backward design is an approach to instructional planning that allows the intended outcomes to guide development of learning and assessment activities, thereby helping students learn what is most essential in the course, program, or discipline. Through this approach, intended learning outcomes become the lodestar that guides the creation and use of summative and formative assessment. Summative assessment evaluates whether students achieve the intended learning outcomes, and formative assessment provides opportunity for practice and ongoing feedback to help them get there. In contrast to instructional design that focuses on the facts and content presented by the instructor, this approach emphasizes student learning. Students in backward-designed courses learn better because the instruction is aligned with outcomes and assessments. Ideally, students can spend more time on activities that help them gain deep conceptual understanding of concepts and give them practice with skills necessary to succeed in science. Thus, backward design helps put an end to the unspoken contract of "we pretend to teach 'em and they pretend to learn" (Wente 2009).

Appendix A: Examples of Alignment between Learning Goals, Intended Learning Outcomes, Summative Assessments, and Learning Activities for Topics from Several Scientific Disciplines

Learning goal	Intended learning outcome	Evidence of achievement (summative assessment)	Learning activity that promotes achievement (formative assessment)
Students will understand gene expression.	Students will be able to predict changes in amino acid sequence that result from changes in the nucleotide sequence of a gene.	Given a known gene and its resultant amino acid sequence, students will use a codon table to predict changes in the amino acid sequence caused by a given mutation in the gene.	Students are given the sequence of a portion of a gene with the corresponding amino acid sequence. Using a codon table, they work in groups to identify the template and coding strands, determine the reading frame, and predict changes in the amino acid sequence that result from a mutation.
Students will understand the relationship between the electronegativity of two atoms and their likelihood to form given types of chemical bonds.	Students will be able to determine the relative electronegativities of two given atoms by using the periodic table and to describe the type of bond the two atoms will form.	Using a periodic table, students will match the correct bond type—ionic, non-polar covalent, or polar covalent—with the pairs of atoms that will form that bond.	Using a periodic table, students will explain the differences in the electronegativities of metals and non-metals and explain why atoms from these groups form different types of bonds based on whether the interactions are between two metals, two non-metals, or a mix of each.
Students will understand the concept of conservation of angular momentum.	Students will be able to predict changes in the speed of spinning objects as they change in size.	When given a scenario such as a collapsing star, students will predict changes in the speed of the star's spin at various points in time.	Students will be given familiar scenarios, such as figure skaters drawing their extended arms to their chests, and asked to predict changes in spin and explain the phenomenon. Then students will identify new examples of angular momentum.
Students will understand the rock cycle.	Students will be able to identify the cause and relative timing of various events that occurred during the building of the Earth's crust.	Using a diagram of a cross section of the Earth's crust depicting various features, students will rank the various features in the order in which they occurred.	Using a diagram of a cross section of the Earth's crust, students will identify and put in order various features caused by sedimentation, weathering, erosion, compaction/burial, deformation/metamorphosis, melting, crystallization of magma, and uplift.
Students will understand how centripetal force and gravity imbalances cause tides.	Students will be able to draw the positions of the Earth-Sun-Moon system that would result in different daily, monthly, or seasonal tides.	Given specific alignments of the Earth-Sun-Moon system, students should be able to draw the solar tidal bulge and the lunar tidal bulge, explaining why their locations differ.	Students work in groups using physical models of the Earth, Sun, and Moon to explain how the positions of these three affect tides at different times of the day, month, and year. Students then draw diagrams of these depictions.

Appendix B: Backward Design in Practice: A Worksheet for Designing and Aligning Instructional Materials

Modified from Tables 5.3 and 5.4 in *Scientific Teaching* (Handelsman *et al.* 2007)

Step 1: Identify intended learning outcomes that explicitly state what students should know and be able to do.			Step 2: Design summative assessments that provide evidence about whether students achieve the outcomes.	Step 3: Design learning activities/formative assessments that help students achieve the outcomes.
Design instruction to address a variety of learning goals for a science course.	Use the spaces below to state broad **goals** with verbs like "understand," "learn," "know," "appreciate."	Use the spaces below to state **outcomes** specifically with verbs like "define," "describe," "explain," "distinguish," "predict," "analyze," "create," "argue," "communicate."	Use the spaces below to design **exam questions** at the same cognitive level as the intended learning outcomes.	Use the spaces below to design **learning activities** at the same cognitive level as the intended learning outcomes.
Knowledge (key terms, basic facts, fundamental concepts, unifying principles, powerful generalizations, and "big ideas" within the discipline)				
Skills (ability to apply the process of science, reason quantitatively, think critically, use models and simulations, and communicate and collaborate effectively)				
Attitudes (curiosity about the world, appreciation of the nature and process of science, and appreciation of the interrelationship of science and society)				

2

Evaluating the Cognitive Levels of Instructional Materials Using an Educational Taxonomy

"What we are classifying is the intended behavior of students—the ways in which individuals are to act, think, or feel as the result of participating in some unit of instruction."

—Benjamin Bloom

During the process of instructional design, an instructor must consider not only whether intended learning outcomes are assessable, but also whether they are aligned with the way students are taught and assessed. Instruction and evaluation are misaligned when the demands on students during class are at different cognitive levels than the demands during examinations. Educational taxonomies provide a straightforward and intuitive framework to categorize cognitive levels of course materials and help instructors check for alignment. This chapter provides a practical method for gauging whether assessments are appropriately aligned with intended learning outcomes and learning activities. By using an educational taxonomy to diagnose and adjust misalignments in their course materials, instructors increase the likelihood that students will achieve the desired levels of learning.

Bloom's Taxonomy

Chapter 1 introduced the use of intended learning outcomes to guide instructional design. In this chapter, we demonstrate the use of a common educational taxonomy, Bloom's Taxonomy, to gauge whether classroom assignments, assessment, and desired learning outcomes are at the same cognitive levels. Bloom's Taxonomy is a simple yet effective tool for evaluating alignment of course materials, thereby increasing the odds that instruction will help students achieve the instructor's expectations for learning.

While there have been many attempts to categorize cognitive levels of understanding (see Appendix B), the most widely adopted is Bloom's Taxonomy, originally published as *The Taxonomy of Educational Objectives, Handbook I: Cognitive Domain* (Bloom 1956) and subsequently adapted by others (e.g., Anderson and Sosniak 1994; Anderson *et al.* 2001; Krathwohl 2002). In discussing Bloom's Taxonomy, we will generally use the modified version by Anderson *et al.* (2001). The taxonomy is a classification system intended to guide curriculum development. It distinguishes six unique categories of human cognition, identified by the verbs *remember, understand, apply, analyze, evaluate,* and *create* (we will subsequently refer to these categories as "Bloom levels").

Often more useful than the distinctions between all six Bloom levels is simply the difference between *higher-order cognitive (HOC)* and *lower-order cognitive (LOC)* skills (Crowe *et al.* 2008). Table 2.1 shows how the six levels can be distilled down to a distinction between HOC and LOC skills and includes a list of more specific action verbs that represent assessment activities associated with each level. We will subsequently refer to these as "Bloom verbs." Hallmarks of LOC skills include memorization as well as recall and explanation (*remember* and *understand*), while typical HOC skills include the breakdown, critiquing, and creation of information (*analyze, evaluate,* and *create*) (Bloom 1956; Zoller 1993; Dickie 2003; Crowe *et al.* 2008). The mid-level *apply* category can be either an LOC or HOC skill, depending on the context of the task. If students were required to use knowledge in a manner that is very similar to how they previously practiced it or how the instructor presented it, the activity or question would be at the LOC level. However, a task that called for students to apply their knowledge to a new situation would be at the HOC level.

The taxonomy is hierarchical with respect to complexity: *remember* is considered the simplest task, and *create* the most complex (Krathwohl 2002). Some educators disagree with the hierarchical structure of Bloom's Taxonomy, argu-

Table 2.1 Bloom's Taxonomy Simplified as Two Cognitive Levels

Cognitive Level	Bloom Level "A simple phrase to guide categorization" Verbs Typically Associated with the Category
HOC	**Create** **"Create something new"** adapt, assemble, compose, construct, create, design, develop, devise, formulate, generate, integrate, invent, make, model, plan, pose, pretend, produce, propose, reconstruct, reframe, revise, rewrite, set up, structure, substitute **Evaluate** **"Defend or judge a concept or idea"** appraise, argue, assess, conclude, criticize, critique, decide, defend, evaluate, judge, justify, prioritize, prove, rank, rate, select, support, validate **Analyze** **"Distinguish parts, organize information, and make inferences"** analyze, break down, categorize, characterize, classify, compare, contrast, correlate, debate, deduce, diagram, differentiate, discriminate, distinguish, examine, infer, outline, question, rearrange, relate, separate, subdivide, test
LOC/HOC	**Apply** **"Use information or concepts in new ways"** act, administer, apply, calculate, change, chart, compute, demonstrate, determine, draw, dramatize, employ, extend, illustrate, implement, inform, instruct, operate, practice, predict, prepare, produce, provide, role-play, show, sketch, solve, transfer, use, utilize
LOC	**Understand** **"Explain information or concepts"** convert, define, describe, demonstrate, discuss, explain, express, generalize, give examples, imitate, indicate, interpret, paraphrase, restate, summarize, translate **Remember** **"Recall information"** choose, count, duplicate, enumerate, find, identify, label, list, locate, match, memorize, name, order, quote, recall, recite, recognize, repeat, report, reproduce, select, sequence, state, tell

ing that students don't always need to remember factual or procedural knowledge to organize information or create something new (Wineburg and Schneider 2009). For example, students can make observations, pose hypotheses, and design experiments without recalling factual information. The "Bloom T" diagram below depicts a partially hierarchical revised Bloom's Taxonomy (Figure 2.1). In the Bloom T version, the *apply* level implies that students need to recall relevant information and determine the meaning of it before using it in a new

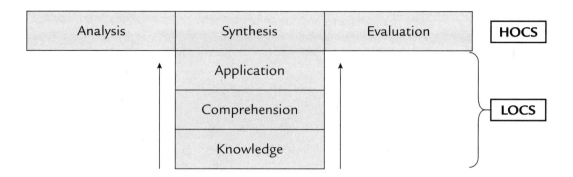

Figure 2.1. A Diagram Showing the Partial Hierarchy of Bloom's Taxonomy

way. However, the HOC levels, such as *analyze*, *evaluate*, or *create*, do not necessarily require factual information and understanding at LOC levels. Moreover, the work of a student at one HOC level does not necessarily depend on their work at another HOC level.

Revised Bloom's Taxonomy in Practice

To evaluate the alignment of course components effectively, an instructor must be able to accurately gauge the Bloom levels of intended outcomes, learning activities, and assessment tasks. This section presents practical guidelines for the categorization of learning materials using the taxonomy.

There are several considerations when categorizing questions and intended learning outcomes at different Bloom levels. First, to classify intended learning outcomes or questions, the appropriate verb should be matched with the Bloom level in Table 2.1. If a verb associated with the required task is not explicitly stated in the table, the instructor will need to infer the verb from what is given or match the task with the phrase provided for each category. If two verbs apply, then the highest category achieved would define the Bloom level. For example, a task requiring students to provide a list (*apply* level) and rank it (*evaluate* level) would be at the *evaluate* level. Appendix C, the Blooming Biology Tool, is designed to help categorize the Bloom level of a question by outlining the types of questions that are typically found at each level. Although the Blooming Biology Tool was based on the original Bloom's Taxonomy, it remains useful because the original and revised taxonomies are very similar.

A second consideration is that Bloom levels are not necessarily measures of difficulty but rather often depend on how the content was taught. Knowledge and comprehension tasks at the lowest two levels, for example, can be relatively difficult when they require retrieval of a large amount of specific information. Likewise, HOC tasks are not always difficult. Posing hypotheses to explain a phenomenon or creating models based on content from textbooks is not always intrinsically difficult. Additionally, the cognitive level of a question may be different for different students depending on their past learning experiences. For example, an *analyze* question for one student may be in the *remember* category for another who previously performed the task in class or as a homework assignment.

A third consideration is the type of assessment question: multiple-choice or free-response. For multiple-choice questions (MCQs), the format of the question and the quality of the distractors (wrong answer choices) determine the Bloom level (see Chapter 5 for details on MCQs). If a question has distractors that are easily eliminated and the answer requires only recall of the content, then the Bloom level would be LOC. In contrast, if the MCQ requires a student to analyze and work through plausible distractors, the Bloom level would be HOC. Both MCQ and free-response questions may contain multiple parts at different Bloom levels. In these cases, Bloom levels are determined for each part independently. If two or more parts are connected, meaning that the student has to correctly answer one part in order to correctly answer another, the overall Bloom level should be defined by the highest-level component.

The last consideration for the overall categorization of a question is whether the question contains discipline-specific science process skills, such as in-depth analysis and interpretation of scientific drawings, diagrams, or figures, or designing an experiment. These kinds of questions are often ranked at HOC levels (see questions 2 and 5 in Chapter 5, Appendix A, for further illustration). To help faculty categorize questions that deal with science process skills, we created an extension of the Blooming Biology Tool (Appendix D) that provides various examples of both general science and biology-specific skill-based questions at different Bloom levels. Although this tool was designed for biologists, the approach can be easily transferred to other Science, Technology, Engineering, and Math (STEM) disciplines.

To help with the categorization of questions at different Bloom levels, we provide a stepwise protocol in Box 2.1. For further instruction on how to categorize different question formats, there are several examples in Appendix A with detailed analysis of how and why each question is ranked at a particular Bloom level.

Box 2.1. A Protocol for Categorizing Questions

1. Classify each question by matching the verb used with a Bloom level. If a verb associated with the task is not found in the question, determine which verb is implied.
2. Review course material to discern if the question was answered directly in class or in the text. If so, the question may simply test recall and would be considered lower-order cognitive (LOC) rather than higher-order cognitive (HOC).
3. Ask the following:
   a. Does the student have to problem-solve (HOC) or merely recall information (LOC)? The more conceptual steps required, the more likely the question is at an HOC level.
   b. If the student has to problem-solve, is the necessary information present so the student must merely apply content knowledge (LOC), or is some information missing so the student must make inferences to arrive at the correct answer (HOC)?
   c. Are you judging the Bloom level based on difficulty of the question, such as giving a difficult question an HOC ranking even if the task is knowledge-based?
4. Use the Bloom T (Figure 2.1) to check off all Bloom levels required to successfully answer the question. Classify the question at the highest Bloom level required.
5. Determine if science process skills are required to answer the question. Refer to the Blooming Biology Tool (Appendices C and D) to help guide your categorization.
6. Identify the question type. If it is a multiple-choice question, are there strong or weak distractors? The more plausible the distractors, the more likely the question is HOC.
7. Determine if the question has multiple parts.
   a. Determine the Bloom level of each part.
   b. Determine if the multi-part question has connected or independent components. If components are connected, categorize the question at the highest Bloom level. If not, the independent components should be treated as separate questions.
8. Ask a colleague or teaching assistant (TA) to categorize the question using this protocol. Compare your results and come to a consensus.

Writing Questions Using Bloom's Taxonomy

Writing a question at each of the six Bloom levels can be particularly helpful in developing the ability to discriminate between the Bloom levels. Before you begin writing questions, identify the concept you would like students to understand. In a stepwise progression, beginning with the LOC skill of *remember*, write questions that represent each level of the cognitive processes in the taxonomy. At each level, check that the question makes sense; ultimately it is the task the student must accomplish that determines the level of a question. Because it is difficult to write MCQs that represent HOC skills (see Chapter 5), we recommend starting with a free-response question and later modifying it into a multiple-choice format. An example of one concept tested at many levels is found in Box 2.2.

Box 2.2. One Concept: Six Bloom Levels

Modified from Crowe *et al.* 2008

Discipline: Physiology

Remember (LOC)
Q: To determine cardiac output for an animal, which two variables do you need to know?

   Requires students to recall information, but not necessarily to understand the significance of each variable.

A: Heart rate and stroke volume determine cardiac output.

Understand (LOC)
Q: In your own words, explain what cardiac output is and why it is significant.

   Requires students to summarize a process described in text or in class in their own words, showing not only an understanding of how various cardiovascular components relate to one another, but also how cardiac output influences other cardiovascular parameters.

A: Cardiac output is the volume of blood pumped by the heart in a given period of time, in liters per minute (L/m). Blood pressure and thus blood flow are determined by cardiac output and total peripheral resistance.

Apply (LOC/HOC)
Q: If cardiac output increases, predict how arterial blood pressure will change. Explain your answer. Lance Armstrong has a normal resting cardiac

output of 6 L/min yet his resting heart rate is only 40 beats/min. What is his stroke volume?

> Both questions require students to predict an effect on a new situation not previously encountered. Neither question is complex; if students know the equations for cardiac output and blood pressure, they should be able to answer the questions correctly.

A: If cardiac output increases, blood pressure should increase unless resistance decreases proportionally. Lance Armstrong's stroke volume is 150 mL/beat.

Analyze (HOC)

Q: Compared to a normal resting male of the same height and weight, Lance Armstrong's stroke volume is greatly increased. Provide a physiological explanation of how he can have such a large stroke volume.

> Requires students to draw inferences from the data.

A: A greater stroke volume is a result of a more forceful contraction of the left ventricle. Because muscle strength is determined by the cross-sectional area of the muscle, Armstrong's left ventricle must be very thick.

Evaluate (HOC)

Q: If a patient's CT scan revealed an enlarged heart, how would you determine if this enlarged heart was pathological or not?

> Requires students to comprehend two different physiological states: highly trained athlete versus disease state. They would have to comprehend the information obtained from different diagnostic tests and determine which would be most informative.

A: Speaking with the patient and getting his history would show that the patient is a highly fit athlete.

Create (HOC)

Q: Create a summary sheet that is a pictorial depiction/flow diagram of how changes in cardiac output influence mean arterial blood pressure. Generate a graph that shows Lance Armstrong's left ventricle volume during the cardiac cycle.

> Requires students to draw inferences from the data and then create a new model that is consistent with the data.

A: Summary sheet would show that if cardiac output increases while resistance is held constant, then mean arterial pressure would increase. Graph would show end diastolic volume of 170 mL that drops to 20 mL for an end systolic volume.

Using Bloom's Taxonomy to Evaluate Instructional Alignment

As previously mentioned, one of the most important uses for Bloom's Taxonomy is to determine if there is alignment between the way students are taught and tested. The following vignette shows how misalignment of a course unit creates frustration for the instructor.

The night before her first exam, an introductory biology instructor finishes writing her exam questions and reviews her intended learning outcomes related to cell cycle, mitosis, and meiosis. She also reflects on how she instructed the students. Here is what she finds:

Intended Learning Outcomes:
1. Students will be able to interpret graphical information from multiple representations (understand, LOC).
2. Students will be able to create graphs related to biological phenomena (create, HOC).
3. Students will be able to list all the steps of cell cycle, mitosis, and meiosis, and describe what happens at each step (recall, LOC).
4. Students will be able to compare mitosis and meiosis (analyze, HOC).
5. Students will be able to predict the consequences of errors in mitosis or meiosis (apply, HOC).

Exam Questions:
1. The graph on the following page shows the results from 1×10^5 cells that have been stained with a DNA-binding dye and analyzed using a flow cytometer. Note: the cells were not synchronized with regard to the phases of the cell cycle and all cells died immediately after adding the dye.
   a) Use an arrow and label "A" where cells in anaphase would be indicated on the graph.
   b) Use an arrow and label "C" to show where cells that have just completed cytokinesis would be indicated on the graph.
   c) If these cells are from an organism that is 2N=4 and a non-disjunction occurred during anaphase of mitosis (a pair of sister chromatids was pulled to one centrosome), then a new population of cells would be detected on the graph. Draw on the graph the new profile of these

cells (assume these cells are approximately one half of the population shown).

(Answers in bold on graph below to the right)

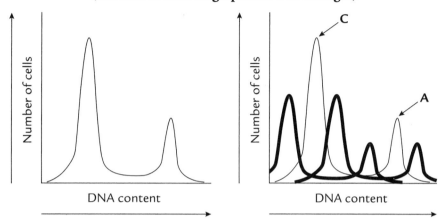

She then realizes the students must interpret a graph and apply their knowledge of mitosis and cell cycle in a new way (apply, HOC). To correctly answer part b, the student must understand the cell cycle, infer that cells that have undergone cytokinesis are in G1, and then compare the relative DNA content in that cell to that of a cell that has finished replicating its chromosomes (analyze, HOC). Part c requires the student to predict the outcome of non-disjunction relative to DNA content and represent it graphically (create, HOC).

2. For an organism having a diploid number of 4 and two loci, F and A, on different chromosomes:
    a. Draw and clearly label a cell with genotype FfAa during metaphase of mitosis.
    b. Draw and clearly label the same cell during metaphase of meiosis I.

Answers:

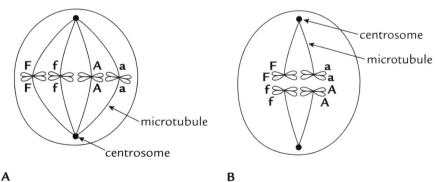

In Question 2, students must be able to differentiate the stages of mitosis from meiosis and represent the different stages pictorially (analyze, HOC).

At this point, she feels that her exam addresses all the intended learning outcomes. However, when she reflects on the alignment between exam questions and how she taught the material, she realizes there is a problem. She only showed her students pictures of the different stages of mitosis and meiosis and never asked them to draw the stages from memory or apply flow cytometry to analyze dividing cells. She worries that, based on the class instruction, students will assume they only have to identify the different stages of mitosis and meiosis, rather than perform these more complex tasks. Acknowledging the misalignment she changes her questions, but wishes that she had thought of this sooner so she could have changed her teaching practices instead.

This vignette highlights the importance of using intended learning outcomes to develop and align assessment and instruction, as discussed in Chapter 1. The Backward Design model encourages creation of exam questions that align with intended learning outcomes *before* instruction begins, so the instructor will know how much and what type of practice to give her students. As a result, she can provide her students with activities that challenge them at relatively high cognitive levels in class so they can deal with challenging questions during exams.

How to Check Alignment of Bloom Levels by Using a Table

A course alignment table like the one shown in Table 2.2 (based on the vignette above) provides a useful overview of alignment between intended learning outcomes, class activities (formative assessments), and examinations (summative assessments) prior to instruction. Once the LOC or HOC levels for the intended learning outcomes, assessment activities, and class activities have been determined, creating a table of the information is straightforward. The instructor in the vignette would have benefitted greatly from this exercise, as she would have immediately recognized that she was not giving her students practice with HOC-level tasks. By creating a table to check alignment of Bloom levels for each teaching unit, instructors can more easily design classroom activities that reflect what they want their students to learn.

Table 2.2. Bloom Levels Alignment Table (for vignette above)

LOC or HOC Level	Bloom Level	Intended Learning Outcomes	Formative Assessments	Summative Assessment
HOC	Create	√	?	√
HOC	Evaluate			
HOC	Analyze	√	?	√
HOC/LOC	Apply	√	√	√
LOC	Understand	√	√	√
LOC	Remember	√	√	√

Conclusion

In today's complex, interdisciplinary world it is increasingly important that students learn to work at relatively high cognitive levels. Indeed, when surveyed, science faculty have indicated that they want students to be proficient at interpreting data, solving problems, critically reading and evaluating different types of literature, communicating science to others, making connections between content areas, and applying scientific content to life (Coil *et al.* 2010). These are the same skills promoted in the NRC *Bio2010* report (National Research Council 2003), *Scientific Foundations for Future Physicians* (Association of American Medical Colleges and Howard Hughes Medical Institute 2009), and the Vision and Change report (Woodin *et al.* 2010).

Without the benefit of an effective learning environment, many students will fail to acquire the knowledge and skills necessary to develop as scientists. Using pedagogical tools such as Backward Design and Bloom's Taxonomy, instructors provide the curricular structure to help students learn and achieve at the highest intellectual levels. While there are other educational taxonomies with which to evaluate alignment of instructional materials, Bloom's Taxonomy is straightforward, easy to use, and flexible. As a consequence, several science educators have modified it for use in their own disciplines (McGuire 1963; Dickie 2003; Pungente and Badger 2003; Lord and Baviskar 2007; Crowe *et al.* 2008), thereby making it accessible to college instructors everywhere.

Appendix A: Sample Questions

The following examples provide detailed analyses of how and why a given question would be ranked at a particular Bloom level. They further illustrate how to categorize questions with varying structure and skill requirements.

Example 1

Type: Free-response, multi-component question with connected components and implied verbs.

Question Level: HOC

Discipline: Introductory Geology—Earth as a System

Reference: Coughenour, C., personal communication

The relatively cold, rigid lithosphere sits atop the more plastic asthenosphere and seeks a condition of equilibrium called isostasy. This phenomenon can be quantified via analysis of buoyant forces:

$$F_{net} = F_g - F_B = m_m g - \rho_{fluid} V_{displaced} g = \rho_m V_m g - \rho_{fluid} V_{displaced} g$$

where F is force, g is gravity, B is buoyancy, and V is volume.

Archimedes found that an object displaces fluid until the mass of displaced fluid equals the mass of the object. Mathematically, we can express the equilibrium condition of floating as:

$$0 = F_{net} = \rho_m V_m g - \rho_{fluid} V_{disp} g$$

In other words:

$$\rho_m V_m g = \rho_{fluid} V_{disp} g$$

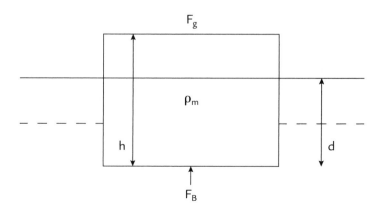

Looking at the volumes in the diagram below, we can now rewrite the above equation as:

$$\rho_m g(lwh) = \rho_{fluid}g(lwd) \text{ or } \rho_m h = \rho_{fluid}d$$

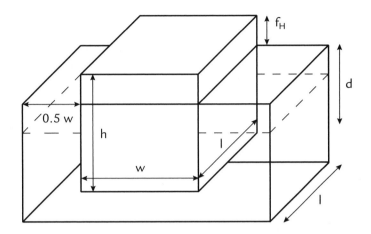

a. How high does a 7-km-thick portion of oceanic crust, which we will take as essentially basalt ($\rho = 3.0$ g/cm³), sit on the mantle ($\rho = 3.3$ g/cm³)? Please give answers in km and show your work. Repeat this calculation for the case of continental crust ($\rho = 2.7$ g/cm³) that is 35 km thick.

b. Calculate the altitude of the continental crust if the top of the oceanic crust is 2,500 m below sea level. Assume that the top of the mantle is stationary and does not change its height.

Answers and Analyses

Answer to part a: To solve for the height (f_H) of the oceanic crust on the mantle, one must solve for d first. Treating the mantle as fluid and the oceanic crust as a solid, we can solve for d using $\rho_m h = \rho_{fluid}d$. Thus, (3.0 g/cm³)(7 km) = (3.3 g/cm³)d, and solving for d gives us 6.4 km. Then we can solve for $f_H = 7$ km – d or $f_H = 7$ km – 6.4 km. Therefore, $f_H = 0.6$ km. Repeating the calculation for the continental crust results in d = 28.6 km and $f_H = 6.4$ km.

Answer to part b: If the oceanic crust sits 0.6 km above the mantle and 2.5 km below the ocean and the continental crust sits 6.4 km above the mantle, then the height of the continental crust above the ocean (Hcc) is: Hcc = 6.4 km – (2.5 km + 0.6 km) or Hcc = 3.3 km.

The first part of this question gives the students the formulas and numbers needed to solve the problems. It does not explicitly use any of the verbs associated with Bloom levels. The question is at the *apply* level and is LOC because the student needs only to apply the information given to perform the required calculations without necessarily needing to have an understanding of geology concepts.

The second part of the question relies on the answers derived from the calculations performed in the first part. Although this second question is also at the *apply* level, it is HOC because the student must use two disparate pieces of information (answers to the calculations for part a) and incorporate new information (that the oceanic crust is 2.5 km below sea level) to solve the problem. Moreover, the student must determine the relationship of all three pieces of information and decide what is being asked.

Example 2

Type: Multiple-choice question with strong distractors

Question Level: LOC or HOC (depending on how the question is written)

Discipline: Introductory Biology

Reference: Freeman 2010

> O_2 will diffuse from blood to tissue faster in response to which of the following conditions?
>
>   a. an increase in the PO_2 of the tissue
>   b. a decrease in the PO_2 of the tissue
>   **c. an increase in the thickness of the capillary wall**
>   d. a decrease in the surface area of the capillary

This question does not explicitly state the verb associated with the Bloom level, but the student must make a selection by comparing potential answers. Thus, the quality of the distractors helps to determine the level at which this question should be categorized. Because surface area, capillary thickness (a barrier to diffusion), and the difference in partial pressure of oxygen (PO_2) on either side of the barrier all impact the rate of diffusion of oxygen, the student would have to analyze each answer and infer how diffusion would change as a result of the condition. Therefore, this would be an HOC question at the *analyze* level.

The same question could be written at a lower level by changing the distractors:

O_2 will diffuse from blood to tissue faster in response to which of the following conditions?

   a. an increase in the PO_2 of the tissue
   b. a decrease in the PO_2 of the tissue
   c. an increase in the PCO_2 of the tissue
   d. a decrease in the PCO_2 of the tissue

In this example, a student only needs to recall that diffusion of a gas is based on the movement of molecules from an area of high concentration to an area of low concentration. The assumption is that the instructor explicitly stated this in class, likely using oxygen or carbon dioxide in the example. In this scenario, the question would be categorized as an *understand*-level (LOC) question.

Example 3

Type: Free-response, multi-component question with connected components

Question Level: HOC

Discipline: Organic Chemistry

Reference: McKinstry, L., personal communication

The thiocyanate ion [SCN]⁻ has the skeletal connectivity indicated here: S-C-N

   a. Draw Lewis diagrams for this molecule, including the possible resonance structures and all formal charges.
   b. Use arrow formalism to show how each resonance contributor is converted to the next.
   c. Rank the resonance contributors in terms of importance in contributing to the overall chemical nature of this compound. Briefly give your reasoning.

Answers and Analyses

Answer to parts a and b:

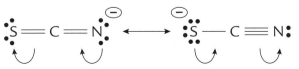

Answer to part c: The structure on the left is more important in contributing to the overall nature of this compound because the negative charge is on the more electronegative atom.

For this question, the different components are connected and the correct answers for parts b and c depend on drawing the correct Lewis diagrams in part a. The different components of the question—a, b, and c—should be categorized separately, but the instructor should use the highest Bloom level for the overall categorization. The verbs "draw," "use," "show," and "rank" are easily categorized (see Table 2.1) and serve as a good starting point for determining which part of the question requires the highest cognitive skill. While "use" and "show" fall under the *apply* category, "draw" is a verb that is found at many levels because drawing a simple representational model is very different from drawing a sophisticated model using little prior knowledge. Drawing Lewis diagrams requires a student to apply numerous rules and therefore is at the *apply* level. Because the student must "rank" the resonance structures, the student has to make a judgment. Therefore, this question would be classified as HOC because the highest cognitive level required is *evaluate*.

Example 4

Type: Free-response, multi-component question with distinct components, requiring science process skills

Question Level: LOC and HOC

Discipline: Interdisciplinary Science (introductory physics, chemistry, and biology)

Reference: Chowdary, K., C. Dirks, and L. McKinstry, personal communication

This diagram represents the first order reaction for $H_2O_2 \rightarrow 2 \cdot OH$:

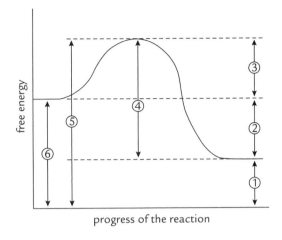

a. Which quantity in the diagram represents the …

activation energy for the forward reaction? _____

free energy of the reactant? _____

change in free energy for the reaction? _____

activation energy for the reverse reaction? _____

b. On the diagram below, carefully draw the free energy curve for the spontaneous reaction with the same reactants and products and the same free energy change, but one that would occur **more quickly in the presence of a catalyst**.

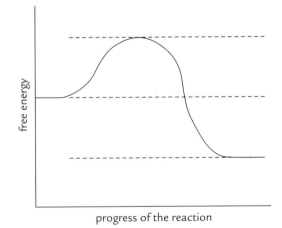

c. Given the rate constant values of k = 1.3 × 10^{-5} s^{-1} at 15 °C and k = 8.0 × 10^{-3} s^{-1} at 50 °C, what is the energy of activation (E$_a$) for the forward process? R is 8.314 J/mol K and T is temperature in Kelvin (K = °C + 273.15).

$$\ln \frac{k_2}{k_1} = -\frac{E_a}{R}\left[\frac{1}{T_2} - \frac{1}{T_1}\right]$$

Answers and Analyses

Answer to part a: 3, 6, 2, 4

Answer to part b: The graph would be identical except that it would show a smaller activation energy for the forward reaction (i.e., arrow 3 would be smaller, showing a decrease in energy).

Answer to part c: Students would need to first convert the temperatures in Celsius to Kelvin and then solve for E$_a$ = [ln (k$_1$/k$_2$)] (8.314 J/mol K)/(1/T$_1$ – 1/T$_2$). Thus, E$_a$ = 90.6 J/mol.

This question requires that students have been introduced to free energy diagrams and have a working knowledge of chemical reactions. It has three parts that are not dependent on each other; students can answer part b or c without having correctly answered part a. Part a is *recall* because students were likely introduced to energy diagrams and shown which part of the graph is the "activation energy." In order for students to answer part b, they need to deduce how the graph can be changed given certain limits (same free energy change), making part b a *create* question. Part c is an *apply*/LOC question because students merely have to plug the values into a given equation (if they had to derive the above equation from the Arrhenius equation, then part c would be a *create* question).

This question also requires skills that are not content-specific: drawing a graph and performing a calculation. While students may be able to recognize and recall the different components of a graph they have already seen, drawing a new graph of a faster reaction, even if they understand what a spontaneous reaction is, may be a daunting task. Sometimes students can explain the outcomes from changing a condition for a particular problem, but fail to translate that into a graphical representation of the answer because they lack basic science process skills. Therefore, when writing exam questions it is important to consider how much practice students have had with the science process skill that makes up part of the question, as this could undermine their ability to answer the question.

Example 5

Type: Free-response, multi-component question with distinct components, requiring science process skills

Question Level: HOC

Discipline: Physics

Reference: Crowell 2000

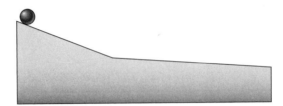

   a. The ball is released at the top of the ramp shown in the figure. Friction is negligible. Use physical reasoning to draw v-t and a-t graphs. Assume that the ball doesn't bounce at the point where the ramp changes slope.

   b. Do the same for the case where the ball is rolled up the slope from the bottom, but doesn't quite have enough speed to make it over the top.

Answers and Analyses

Answer to part a: We choose a coordinate system with positive pointing to the right. Some people might expect that the ball would slow down once it were on the less steep ramp. This could be true if there were significant friction, but Galileo's experiments with inclined planes showed that when friction is negligible, a ball rolling on a ramp has constant acceleration, not constant speed. The speed doesn't increase as quickly once the ball is on the less steep slope, but it still keeps increasing. The a-t graph can be drawn by inspecting the slope of the v-t graph.

Answer to part b: The ball will roll back down, so the second half of the motion is the same as in part a. In the first (rising) half of the motion, the velocity is negative, since the motion is in the opposite direction compared to the positive x-axis. The acceleration is again found by inspecting the slope of the v-t graph.

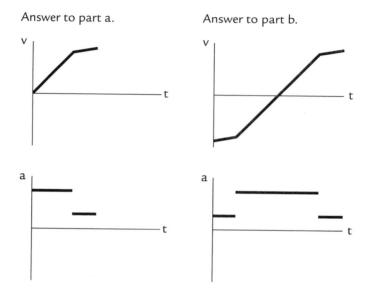

This question requires students to analyze the diagram, infer what will happen under the conditions stated, and show the results of their inferences as graphs. Both parts are at the HOC level (*analyze*) but are not connected because the answer to part b does not rely on the answer to part a. Interestingly, the answer to part b is not just a reversal of the answer to a; therefore, the students must truly analyze both situations to arrive at the answers.

Appendix B: A Summary of the Evolution of Seminal Taxonomies in Education

Instrument	Description	Authors & Year
Bloom's Taxonomy	Three domains related to learning: • Cognitive: knowledge-acquisition skills • Affective: feelings, beliefs, and attitude • Psychomotor: physical skills	Bloom 1956
Revised Bloom's Taxonomy	Bloom's Taxonomy modified with action verbs instead of nouns, reordering of Bloom levels (synthesis and evaluation), and going from one- to two-dimensional analysis	Anderson et al. 2001
Structure of Observed Learning Outcomes	Categorizes learning from a surface to deep approach	Biggs and Collins 1982
Table of Learning	A non-hierarchal taxonomy incorporating many areas of student cognition	Schulman 2002
The New Taxonomy of Educational Objectives	Integration of the three domains of learning: cognitive, affective, and psychomotor	Marzano and Kendall 2008

Appendix C: Blooming Biology Tool

	Knowledge[1]	Comprehension[1]	Application[1]		Analysis	Synthesis	Evaluation
	LOCS[2]	LOCS[2]	LOCS[2]	HOCS[3]	HOCS[3]	HOCS[3]	HOCS[3]
Key Skills Assessed	**IDENTIFY, RECALL**, list, recognize, or label	**DESCRIBE** or explain in your own words, re-tell, or summarize	**PREDICT** an outcome using several pieces of information or concepts; use information in a new context		**INFER**; understand how components relate to each other and to the process as a whole	**CREATE** something new using/ combining disparate sources of information	**DETERMINE/ CRITIQUE** relative value; determine merit
General Examples of Biology Exam Questions	Identify the parts of a eukaryotic cell; identify the correct definition of osmosis	Describe nuclear transport to a lay person; provide an example of a cell signaling pathway	Predict what happens to X if Y increases		Interpret data, graphs, or figures; make a diagnosis or analyze a case study; compare/contrast information	Develop a hypothesis; design an experiment; create a model	Critique an experimental design or a research proposal; appraise data in support of a hypothesis
Types of Exam Questions							
labeling	X	X	X				
fill-in-the-blank	X	X	X		X		
true-false	X	X	X		X		
multiple-choice	X	X	X		X		X
short answer	X	X	X		X	X	X
essay	X	X	X		X	X	X
Characteristics of Multiple-Choice Questions	Question requires only information recall; possible answers do not include significant distractors[4]	Question requires understanding of concept or terms; possible answers include significant distractors[4]	Question requires prediction of the most likely outcome given a new situation or perturbation to the system		Question requires interpretation of data and selection of best conclusion	N/A: If provided with choices, students only differentiate between possible answers rather than synthesize a novel response	Question requires assessment of information relative to its support of an argument

[1]The first three Bloom levels are usually hierarchical; thus, to complete an *analysis* level question, students must also demonstrate *knowledge, comprehension,* and *application* level skills.

[2]LOCS, Lower-Order Cognitive Skills

[3]HOCS, Higher-Order Cognitive Skills

[4]Significant distractors are those answers that represent common student misconceptions on that topic.

Appendix D: Extension of the Blooming Biology Table
Examples and Descriptions of Science-Specific Skills at Different Levels of Bloom's Taxonomy

	Knowledge[1]	Comprehension[1]	Application[1]		Analysis	Synthesis	Evaluation
	LOCS[2]	LOCS[2]	LOCS[2]	HOCS[3]	HOCS[3]	HOCS[3]	HOCS[3]
Calculations	Equation provided and variables identified; "plug and chug"	Understand/define components and variables of a given equation	Solve word problems by selecting correct formula and identifying appropriate variables	Solve word problem and infer biological significance or implication	Create an equation that describes the relationship between variables	Evaluate a computational solution to a problem or assess the relative merit(s) of using a specific mathematical tool to solve a particular problem
Concept Maps	Structure provided; student fills in the missing linking phrases or concepts that are provided	Structure provided with concepts filled in; student generates linking phrases to describe relationships	Student creates the structure; concepts and linking phrases provided	Student creates structure; concepts are provided; student generates linking phrases to describe relationships and must link two different domains or maps together	Student creates structure and generates concepts and linking terms; map must be sufficiently complex	Student evaluates existing concept maps based on established criteria/rubric
Diagnoses	Identify or list variables found in patient history, vital signs, and/or clinical test results; know which physiological problem each named disease represents (e.g., Graves' Disease, hyperthyroidism)	Define each variable; define the presenting signs and symptoms of each disease	Given a set of clinical variables, identify the relevant variables and make a diagnosis	Given a set of clinical variables and a diagnosis, determine which other possible diseases (differential diagnoses) need to be ruled out	Given a set of clinical variables and a diagnosis, determine the next clinical test that needs to be performed to confirm the diagnosis	Given a set of clinical variables and a diagnosis, evaluate the evidence supporting the diagnosis and provide the patient with a second opinion

(continued on next two pages)

	Knowledge[1]	Comprehension[1]	Application[1]		Analysis	Synthesis	Evaluation
	LOCS[2]	LOCS[2]	LOCS[2]	HOCS[3]	HOCS[3]	HOCS[3]	HOCS[3]
Graphing	Identify the parts of graphs and recognize different types of graphs (e.g., identify the X axis; identify a histogram)	Describe the data represented in a simple graph	Draw a graph based on a given set of data; predict outcomes based on data presented in graph		Read and interpret a complex graph having multiple variables or treatments and explain biological implications of data	Create a graphical representation of a given biological process or concept	Assess the relative effectiveness of different graphical representations of the same data or biological concept
Hardy-Weinberg Analyses	Given the Hardy-Weinberg (HW) equation, define terms: p^2, q^2, $2pq$; if given $p + q = 1$ and $p = 0.7$, calculate q; if given that the alleles p and q in a population equal 1, and $p = 0.7$, identify the recessive allele and calculate q	Describe the assumptions of the Hardy-Weinberg equation and its use as a null hypothesis; explain what $2pq$ represents in the Hardy-Weinberg equation (HW equation not given)	If the recessive allele is represented in 30 percent of the population, determine the expected number of homozygous recessive individuals in that population (HW equation not given)		Based on a sample of 100 individuals from a population, in which 37 are SS, 8 are ss, and 55 are Ss, determine if this population is in Hardy-Weinberg equilibrium and explain why or why not	Create a new version of the Hardy-Weinberg equation that incorporates 3 alleles	Student would analyze chi-square analyses to weigh predicted evolutionary flux
Molecular Techniques	Identify what is being measured by a molecular technique (e.g., Northern analysis measures relative RNA levels in a given cell or tissue)	Understand what the results of a molecular technique indicate (e.g., the intensity of a band on a Northern blot indicates relative expression of a specific mRNA in the cell type or tissue from which the RNA was obtained)	Draw the expected results you would obtain from a given molecular technique or state which technique could be used to solve a novel problem (e.g., draw the banding pattern you would expect if you analyzed a protein complex containing a 55 kDa protein and a 35 kDa protein by SDS-PAGE)		Interpret the raw data obtained from a molecular technique, including the interpretation of controls and how to normalize data (e.g., interpret the results of a RT-PCR gel analysis by comparing relative expression of experimental genes to a standardized control gene)	Design an experiment using a given molecular technique to test a hypothesis (e.g., design an experiment using Northern analysis to test the hypothesis that transcription factor A regulates expression of gene B)	Assess relative merit of using two different molecular approaches to address a particular hypothesis (e.g., discuss the relative merits of using chromatin immunoprecipitation versus electrophoretic mobility shift assay to test the hypothesis that a protein binds directly to the promoter of a particular gene)

	Knowledge[1]	Comprehension[1]	Application[1]		Analysis	Synthesis	Evaluation
	LOCS[2]	LOCS[2]	LOCS[2]	HOCS[3]	HOCS[3]	HOCS[3]	HOCS[3]
Phylogenetic Tree/ Cladogram	Given a clado-gram, circle the root, nodes, or monophyletic groups	Describe the relationship of sister taxa in a cladogram	Given four clado-grams, identify which one is different and describe the evo-lutionary relation-ships that make it different		If all taxa except for one in a pictured tree have a given synapo-morphy, what can you infer about the evolutionary history of a particular taxa? With respect to that same synapomor-phy, what can you conclude about the most recent common ancestor of all the taxa?	Given a variety of syn-apomorphies from different organisms, create a cladogram, identifying where the derived shared char-acteristics were acquired	Given a case study showing that a group of organisms have different relationships depending on the type of data used to construct the tree, use new information provided to evaluate the collective data and infer the best true relationship of the organ-isms
Punnett Squares and Pedigree Analyses	Punnett square pro-vided; student identifies components (genotypes or phenotypes, parents or offspring) of a given genetic cross	Parental geno-types provided; student makes own Punnett square to show or describe offspring's genotypes and phenotypes	Parental geno-types provided in a word problem; student identifies variables and makes Punnett square to deter-mine genotypic or phenotypic ratios of offspring. Stu-dent is provided with informa-tion regarding dominance, sex linkage, cross-ing-over, etc.		Parental genotypes provided; student makes Pun-nett square to show or describe off-spring's geno-types and phenotypes, and then solves a word problem with the new infor-mation. Student must infer relationships regarding dominance, sex linkage, crossing-over, etc.	Student uses pedigree analysis to develop a hypothesis for how a certain disease is transmitted	Student weighs the relative value of different pieces of evidence (ped-igree chart, incomplete transmission, linkage anal-ysis, etc.) and determines the proba-bility that an individual will develop a cer-tain disease

[1]The first three Bloom levels are usually hierarchical; thus, to complete an *analysis* level question, students must also demonstrate *knowledge*, *comprehension*, and *application* level skills.

[2]LOCS, Lower-Order Cognitive Skills

[3]HOCS, Higher-Order Cognitive Skills

[4]Significant distractors are those answers that represent common student misconceptions on that topic.

II

ASSESSMENT IN PRACTICE

The chapters in this section build upon the theoretical foundation from Section I, offering practical methods and tools to incorporate effective assessment into the college science classroom. The first two chapters elucidate the roles of summative and formative assessment and present approaches for their successful implementation to both evaluate and facilitate learning. The third chapter provides detailed instruction for the creation of challenging assessment questions that can be used in summative or formative assessments. The final chapter introduces students to the cognitive science behind effective study methods.

3

Summative Assessment

Assessment *of* Learning

Show me what is assessed and I will show you what students learn.

—Anonymous

For many instructors, the type of summative assessment they use is based on habit or a method they experienced when they were students. It would seem that we not only teach the way we were taught, but also test the way we were tested! Our traditional methods of assessment, such as multiple-choice exams, are for the most part convenient, and their scores give an illusion of being scientific because they are precise and objective. This quantitative approach can be comforting to science faculty, and gives them a genuine belief in the assessment method (Biggs and Tang 2009). However, the shift from a teaching-centered to a learning-centered classroom (Barr and Tagg 1995), supported by findings in cognitive science and science education research, also calls for a shift in summative assessment from a method that only audits student learning, to a tool that enhances student learning. This chapter discusses the different types of summative assessment and how to use each to enhance student achievement of intended learning outcomes.

Summative Assessment and Student Learning

Student study strategies during preparation for an exam can generally be characterized as either "superficial" or "deep." Students may change these strategies from one course to the next based on their perceptions of course demands (Marton and Saljo 1976; Biggs 1979; Entwistle and Ramsden 1982; Marton *et al.* 1984; Ramsden 1992; Biggs 1993). Courses that encourage a deep approach to learning provide practice and support for the kinds of learning expected, include explicit instruction in the type of mental processes required to succeed, and place emphasis on the student becoming an independent learner. In contrast, courses that tend to encourage a superficial learning strategy usually cover a large amount of material and are perceived to have a threatening and anxiety-provoking assessment system (Gibbs 1992; Kember and Leung 2008). Courses in the Science, Technology, Engineering, and Math (STEM) disciplines often fall into this latter category; therefore, science instructors in particular should pay attention to the forms of summative assessments they use.

Summative Assessment in Practice

The main difference between summative and formative assessment is their intended purpose. The former is assessment *of* student learning, while the latter is assessment *for* student learning (see Chapter 4 for further discussion of the latter). Summative assessment concerns the product, while formative assessment concerns the process. Summative assessment is about accountability, validation, and accreditation, while formative assessment is about enhancement and facilitation of learning.

Summative assessment usually takes the form of an exam but can also be a final report, presentation, or project at the end of a course or a unit of study (Table 3.1). By its very nature, summative assessment is administered periodically, and often the only feedback the student receives is a score, since many faculty want to reuse their exam questions and so do not return graded exams. Major criticisms of summative assessment include the time pressures for taking the test, lack of feedback to inform students about what they are doing wrong and how to correct it, and the tendency of students to focus on the assessment grade rather than mastery of the material. Moreover, because summative assessment is often limited in scope or is poorly aligned with instruction (see Chapters

2 and 4), it may not accurately assess a student's level of mastery of the topic. As shown in Table 3.1, there are trade-offs with using different kinds of summative assessment; the trade-offs usually revolve around time and energy for both student and instructor.

Table 3.1. Advantages and Disadvantages of Several Forms of Summative Assessment

Modified from P. Race (Race 1996)

Assessment	Advantages	Disadvantages
Multiple-Choice Exams	• Students can be anonymous* • More content can be tested • Grading is easy • Grading is objective	• Students can game the test • Students can guess answers • Not possible for students to earn partial credit • Writing HOC MCQs is labor-intensive for instructor
Free-Response Exams	• Students can express themselves in their own words • Students can display creativity • Students can demonstrate higher cognitive skills • Students can be anonymous* • Potential to shift exam questions from lower to higher Bloom levels	• Students usually receive little to no timely feedback • Test format may pose time constraints for students • Students' poor writing skills can detract from answer • Grading is subjective, as it is based on interpretation of written answers • Grading is labor-intensive for instructor
Essays	• Students can demonstrate depth of understanding of a topic • Students can express themselves in their own words • Students can display creativity • Students can be anonymous* • Potential to shift exam questions from lower to higher Bloom levels	• Students' poor writing skills can detract from answer • Test format may pose time constraints for students • Grading is labor-intensive for instructor • Grading is subjective, as it is based on interpretation of written answers
Oral Exams	• Students can fully express their ideas • Students with strong oral communication skills can showcase this talent • Nearly impossible for students to cheat	• Students may be intimidated by speaking to instructor • Students are not anonymous • Students with weak communication skills may be negatively impacted • Only a limited number of topics can be tested in allotted time • Evaluation is subjective unless instructor uses a validated and reliable rubric
Practical Exams	• Students can demonstrate workplace-related skills • Can provide an opportunity for students with greater manual dexterity to showcase this skill	• Students may be nervous about being observed • Students are not anonymous • Can be difficult for instructor to assess objectively without a valid and reliable rubric • Time-consuming for instructor to set up

*Anonymity is possible if student names are concealed or replaced with a unique identifier, thereby avoiding stereotype bias in grading

(continued on next page)

Table 3.1. Advantages and Disadvantages of Several Forms of Summative Assessment (cont.)

Modified from P. Race (Race 1996)

Assessment	Advantages	Disadvantages
Open-Book or Open-Notes Exams	• Potential to shift exam questions from lower to higher Bloom levels • Measure students' ability to find and use information • Students can be anonymous*	• Require more desk space for students' books or notes • May give students a false sense of security and discourage adequate test preparation • Students may have difficulty with time management during the test
Presentations	• Students practice professionalism in the discipline • May facilitate collaborative effort by students • Students can display creativity • Can provide students an opportunity to synthesize a body of work	• Time-consuming for students to prepare • Students may be intimidated by speaking to instructor • Some students dislike public speaking • Students may be nervous about being observed • Students are not anonymous • Can be difficult to discern individual student contribution in a group presentation • Evaluation is subjective unless instructor uses a validated and reliable rubric
Written Reports	• Students have ample time to complete assignment if sufficient lead time is given • Can provide students an opportunity to synthesize a body of work • Can provide practice in a valuable workplace skill • May facilitate collaborative effort by students • Students can demonstrate depth of understanding of a topic • Students can express themselves in their own words • Students can display creativity • Students can be anonymous*	• Labor-intensive for students to prepare • Have a greater potential for plagiarism • Grading is labor-intensive for instructor • Evaluation is subjective unless instructor uses a validated and reliable rubric
Portfolios	• Students can demonstrate maturation of work over a period of time • Can provide opportunity for creative expression by students • Allow students' work to take multiple formats	• Students are often unfamiliar with the format of a portfolio • Portfolios often require special software to create, modify, display, and store material • Evaluation of portfolios is labor-intensive for instructor • Evaluation is subjective

*Anonymity is possible if student names are concealed or replaced with a unique identifier, thereby avoiding stereotype bias in grading

Although summative assessments encompass a wide range of formats, each with its strengths and shortcomings, the most common are multiple-choice and free-response exams. Therefore, we will discuss these types of exams in more depth.

Multiple-Choice Exams

The major advantages of multiple-choice exams are that the grading is objective, the answer sheet is readily graded by machine, and results are easily subjected to item (question) analysis. Item analysis reveals the difficulty and discrimination power of each question; based on this information, instructors can eliminate misleading items from the exam and improve items that will be used on later tests. Because it is less time-consuming to both answer and grade multiple-choice questions (MCQs), it is possible to have a greater number of questions on the exam and for the exam to cover more course material. The increased number of questions and greater scope of content on an exam can improve both the overall validity (the test is assessing what it is intended to assess) and reliability (the test is a consistent measure over time) of the exam (see Chapter 7 for further explanation of item analysis, validity, and reliability).

Because test-taking has been shown to enhance learning, multiple-choice exams have the potential to enhance student performance on subsequent exams. This finding, called the "testing effect," is discussed in greater detail in Chapter 6 (Marsh *et al.* 2007). Unfortunately, multiple-choice exams are a double-edged sword in this regard because students have been shown to recall information from the incorrect distractors as true on subsequent exams. This has been termed the "negative testing effect," as the testing has helped students learn false information. It is thought that the time students spend on reasoning through each possible choice can strengthen and reinforce faulty reasoning (Roediger and Marsh 2005; Fazio *et al.* 2010). This is especially troubling because as class sizes grow and instructional support diminishes, there is increased pressure to give multiple-choice exams to save grading time.

An additional concern with multiple-choice exams is that students often employ surface-learning approaches when studying for such exams (Scouller 1998). A survey of education students on their perceptions of multiple-choice and essay exams given in the same course showed that students saw multiple-choice exams as testing only facts (ibid.). These same students indicated that they were more likely to employ surface-learning approaches when studying for these tests, and felt that a deep-learning approach could lead to a poor performance on multiple-choice exams. On the other hand, these students perceived that free-response and essay exams required higher-order cognitive (HOC) thinking, and they therefore shifted their studying to a deep-learning approach. Students also attributed a poor performance on essay exams to the use of surface-learning study methods (ibid.).

Free-Response Exams

Free-response exams employ open-ended questions that require the student to generate an answer rather than merely recognize a correct option from a set of distractors. This type of question more readily lends itself to testing the HOC skills of analysis, synthesis, and evaluation (see Chapter 2). Free-response questions provide students with the opportunity to express themselves in their own words and to demonstrate the depth of their understanding of the subject. For the instructor, students' written responses often provide insight as to which misconceptions students still hold on the topic. However, there are some major disadvantages of free-response exams, such as the labor-intensive grading for the instructor, the limited number of topics that can be tested at any one time, the time constraints students experience as they formulate and write out the answer, and the impact the students' writing skills have on the quality of the answers.

Two issues arise when instructors grade free-response questions: subjective grading and stereotype bias. Because instructors must interpret students' written answers, grading free-response questions is much more subjective than grading MCQs. However, use of a well-developed rubric (see following section) can help to minimize this subjective bias. To avoid stereotype bias, all answers should be graded blind by concealing or replacing students' names on the exam with unique identifiers. Currently, great advances are being made in machine grading of free-response or constructed response answers through lexical analysis (Ha *et al.* 2011). The availability of this type of grading tool will eliminate many (though not all) of the disadvantages of free-response assessments, foremost being the labor-intensive demands of grading for faculty teaching large courses.

Grading Summative Assessment

Rubrics

A rubric is a scoring tool used on free-response or essay-type assessments that specifies the criteria and level of performance expected by the instructor. When this rubric is shared with students it also makes expectations and scoring explicit (Walvoord and Anderson 1998; Allen and Tanner 2006). Rubrics have been used to assess students' scientific writing skills (Timmerman *et al.* 2010), guide grant proposal construction in an independent-study cell biology laboratory (Crowe *et al.* 2008), grade work in inquiry laboratories (Lunsford and Melear

2004), and efficiently grade laboratory notebooks (C. Dirks, personal communication).

Rubrics can also be learning tools, since they give structure to cognitive activities, provide meaningful feedback, encourage reflective practice, and can help students become self-directed learners (Luft 1999). A rubric can be most useful when it is given to students prior to the assignment as a guide for completing their work. Learning scientists have found that a rubric facilitates both acquisition and retrieval of knowledge for students by providing them with a conceptual framework (Bjork 1979; deWinstanley and Bjork 2002). By providing a framework, the rubric guides a novice through the steps an expert would use to formulate an answer or an assignment (Bresciani *et al.* 2004). Rubrics can also be used to facilitate peer grading (see following section). Some instructors have found that when students use a rubric to grade each others' work, they gain a better understanding of the expectations (intended learning outcomes) of the course (Timmerman and Strickland 2009).

The steps for creating a rubric include the following:

1. Review the intended learning outcomes for the unit, paying particular attention to the cognitive level at which students are expected to perform.
2. Determine the type of evidence that will be accepted to demonstrate mastery by the students.
3. Create a grid showing different levels of completeness of students' answers in each column. Using more than three criterion levels requires the grader to better differentiate between student performances that fall in the mid-range of answers.
4. For each level, clearly define the number or quality of elements required to earn points for that level. The more specific the rubric is for each gradation of points earned, the more useful the rubric will be in grading.
5. Implement the rubric on a subset of answers and modify the rubric as needed.

Rubrics have been classified as either general (holistic) or analytic, depending on the amount of detail they contain (Allen and Tanner 2006). General rubrics, as their name implies, supply only a general sense of what elements should be present in the answer to earn varying levels of points. Analytic rubrics are more labor-intensive to generate but provide both the grader and the student with specific examples that are appropriate for the given assignment. Examples of each type of rubric can be found in Appendix A, and Appendix B provides examples of the numerous websites that offer guidelines and resources for rubric development.

Peer-Grading

Another option for grading free-response exams or written assignments is peer review: having students grade each other's work using a rubric provided by the instructor. However, it is imperative that the process be consistent and rigorous. Students grading peers' work tend to be more lenient than instructors, and are more accurate graders on questions assessing LOC thinking than those testing HOC thinking (Freeman and Parks 2010). One way to decrease the variability associated with peer grading is to use the web-based Calibrated Peer Review (CPR)™ program that provides a structured and systematic method for calibrating a student's ability to evaluate a set of standardized answers. A given calibrated grading ability is weighted in a formula used to normalize a student's review of peer's work (Robertson 2001). The CPR™ method has been used effectively in assessing and improving writing assignments in physiology (Pelaez 2002), biochemistry (Hartberg *et al.* 2008), and other disciplines. More information on CPR™ is found in Appendix C.

Using Summative Assessment as a Learning Tool

Incorporating reflection and feedback into the learning process is critical to promoting a deep approach to learning and to giving students a more realistic understanding of what they know and how well they know it. Summative assessment, unlike formative assessment, often does not provide students with either timely feedback on their errors or an opportunity to correct them. However, it is possible to transform summative assessments into error detection devices that enhance and encourage deep learning (Williams *et al.* 2011). Some strategies for doing this are listed in Table 3.2 and discussed in the following section. Note that each of the methods incorporates elements identified by cognitive science as essential to student learning: opportunities to practice (Balch 1998), encouragement to thoroughly consider all answer options (McClain 1983), immediate feedback for error correction (Friedman 1987; Sadler 1989), and opportunities to monitor their learning (metacognition) (Bransford *et al.* 1999).

Table 3.2. Summative Assessment as a Learning Tool

During the Test

    Collaborative test-taking

    Pyramid exams

    Immediate Feedback Assessment Technique (IF-AT)

    Self-corrected exams

Prior to the Return of the Test

    Do-over

After the Return of Test

    Highlighting missed material

    Point-recapture

    Test analysis

During the Test

Collaborative Test-Taking

Collaborative skills are highly valued and sought-after in the workplace and have been shown to enhance learning when used in the classroom setting (Stearns 1996; Yuretich *et al.* 2001). Collaborative learning (Smith and MacGregor 1992) is based on Vygotsky's views that knowledge is socially constructed and learning is rooted in social interactions (Vygotsky and Cole 1978; Vygotsky 1981). Collaborative exams can be valid, reliable, efficient, and effective, and in a study students found them to be both fair and more valid than individual exams (Shindler 2004). Collaborative tests also had the unintended benefit of increasing students' motivation for studying and critical thinking. Similar results were seen in a Research Methods and Statistics course when students first took a test individually and then with a small group of students, after which the group graded the group exam and discussed incorrect answers (Stearns 1996). Students' scores were based on an average of the individual and group score. Students showed greater retention of material and enjoyment of class—and engaged in more lively discussions while seeking correct answers—when working in a group. However, collaborative assessment is seldom used in higher education.

One issue to consider when doing group work is the composition of the group. Most often instructors allow students to form their own groups, which can lead to uneven distribution of academic talent or the exclusion of quiet students or those for whom English is a second language (ESL). To avoid these pitfalls, instructors can form either homogenous or heterogenous groups based on performance on a standardized test (Lawson 1978) or concept inventory (see Chapter 7 for more on concept inventories). Research on the effect of group composition on student performance shows that students at high risk of failing do better on exams that test at higher cognitive levels when they work in a homogenous group (Jensen and Lawson 2011).

Pyramid Exam

The pyramid exam is a more elaborate form of collaborative test-taking that can take multiple sessions to finish and is used to assess higher order cognitive skills on very complex problems (Cohen and Henle 1995). Students take the test individually for a given period of time for a grade. During the same class period, they then form groups and work together on the exam without the aid of books or notes. At the end of the group session, the "second effort" exam is turned in for the second portion of the grade. Outside of class, students continue to work on the problems individually using any outside aids they wish to, and then turn this exam in at the start of the next class for another individual grade. The fourth stage of the exam takes place in the next class as students work together to write one report. The report is submitted as the group's effort and each member of the group receives the same grade. The following summarizes the four steps of a pyramid exam:

Step 1	In class, alone	(no aids – 50min)	individual grade	20–50%
Step 2	In class, group	(no aids – 50 min)	group grade	20–50%
Step 3	Out of class, alone	(aids allowed)	individual grade	20–40%
Step 4	In class, group report	(no aids – 50 min)	group grade	20–40%

Immediate Feedback Assessment Technique (IF-AT)

Multiple-choice exams generally do not provide corrective feedback that could facilitate learning and retention (Epstein *et al.* 2001; Epstein *et al.* 2002). To increase the learning effectiveness of these exams, Epstein and colleagues created a new testing procedure in which students answer a question until they get the correct answer. This immediate feedback assessment technique (IF-AT) employs a "scratch-off" form for answering multiple-choice questions. Students determine what they think is the correct answer and then scratch-off the thin opaque film covering the answer option. If their selection is correct an asterisk is revealed. Students earn diminishing credit as more attempts are made. Students using the IF-AT forms were found to be actively engaged in the discovery process, and this engagement promoted correction of initially inaccurate ideas and greater long-term retention of material learned. When using IF-AT in conjunction with team-based learning in an Introductory Biology course, students showed considerable learning gains compared to when the course was taught in a traditional lecture format using conventional summative assessment methods (Carmichael 2009). More information on IF-AT is found in Appendix C.

Self-Corrected Exams

In the self-correcting method, students prepare for and take a multiple-choice exam in class, writing their answers in ink rather than pencil (Montepare 2005). They then take the exam home and review their answers using any and all resources at their disposal, inclusive of discussions with other students in the class. Students are allowed to change their answers by crossing out their original choice and indicating the new one; at the next class they turn in the "self-corrected" exam. Each correct answer with no change receives full credit, each corrected answer receives half-credit, and all incorrect answers—originally wrong and unchanged, or changed to wrong—receive no credit.

Prior to Returning the Test

Do-Over

After the instructor has graded a free-response exam, but before it is returned to the student, the instructor selects questions on which the class as a whole did poorly and gives these questions to the class to "do-over." In this scenario, students have not seen the key and are unaware if the answer they wrote is correct, partially correct, or wrong.

The do-over can be done using any one of the following three formats:

1. students work in groups in or out of class and hand in one set of questions to be graded (decreases instructor workload);
2. students consult with classmates in or out of class, but then write and turn in their own answers (improves student metacognition by providing the opportunity to gauge their own understanding compared to their peers' understanding);
3. students are given the questions to take home and complete individually using whatever study materials they wish.

The instructor predetermines the percentage of points that students can earn in each of these formats, and in all cases the student's score on the do-over is added to—rather than replacing—the original score from the exam. Thus, a student who gave a correct answer on the exam and had enough confidence in that answer to submit it again in the do-over could earn twice the points on that question. As the do-over is completed prior to returning the exams, this exercise helps students readjust expectations of their exam performance, making the actual return of exams less traumatic for everyone involved.

After Returning the Test

Highlighting Missed Material

After an exam, students often come to office hours wondering why they didn't receive full credit for their free-response answer, which they feel is no different from the key. To help students better determine how their exam answer differs from the key, students are instructed to highlight on the key all material not found in their answer. Based on what is highlighted on the key, the student then writes a one-page analysis of their exam performance indicating the three major changes they will make to their studying and test-taking strategies for the next exam. The analysis helps the students themselves determine how to improve their performance, rather than hearing an edict from the instructor on the need to change their study patterns. This procedure also has the benefit of making post-exam office hours less argumentative and more productive as students quickly realize how much they actually missed on the exam.

Point-Recapture

The point-recapture method allows students to recapture a percentage of missed points by submitting a written report that includes 1) the right answers, 2) the source for the right answers, 3) a description of why the original answers lost credit, and 4) a plan to avoid similar mistakes on the final exam (Martin-Morris and Wright 2007). This procedure has been shown to minimize grade complaints, encourage better retention of material, and allow poor test-takers an opportunity to improve test-taking skills while avoiding the consequences of a low test score.

Test Analysis

As in the point-recapture method, the instructor corrects the exam and indicates where students received partial credit and where there were errors. After the test has been returned, it is up to the student to correct these errors. The students' final exam scores reflect their responses to the test analysis as well as the initial answers (Carter 1998). This format shifts the responsibility for earning credit to the students, who are encouraged to work together to identify and correct their errors (ibid.).

Test analysis and the do-over differ in the level of feedback students receive prior to resubmitting the exam. In test analysis, the students have received feedback from the instructor as to what they got right and wrong prior to submitting the error corrections. In the do-over method, students are only informed that the class as a whole performed poorly on the do-over questions, but individual students have no indication of how they performed on each question.

Conclusion

Summative assessment is a powerful driver of student learning because students are willing to put in substantial study time to get the score or grade that may help them toward their career aspirations (Morgan *et al.* 2007; Wormald *et al.* 2009). However, students fear the high-risk nature of exams (Anderson 2007) and often do not have a good sense of what to expect or how to study for them (Morgan *et al.* 2007). Since students study what they think will be on the exam, assessment often determines what and how students study more than the actual curriculum does (Morgan *et al.* 2007; Wormald *et al.* 2009). Given that

Table 3.3. Key Elements of Summative Assessment

Summative assessment supports and enhances learning when . . .

it is well aligned with intended learning outcomes;

a range of assessment methods are used to reflect the diversity of learners in the classroom;

classroom learning activities have been designed to give students adequate practice at the appropriate cognitive level;

students receive timely feedback and opportunities to reflect on their understanding;

it is used to diagnose student learning challenges and inform changes in instruction;

there is a shift in emphasis from grades to mastery of material.

assessment is often a very time-consuming task for instructors and is the major determinant of whether a student adopts a deep or surface approach to learning, creating effective and efficient assessment is worthwhile for both the instructor and the student. Summative assessment, therefore, has the potential to be an influential tool for promoting learning. In closing, Table 3.3 summarizes the main points of summative assessment.

Appendix A: Types of Rubrics

I. General (Holistic) Rubric

0 pts.	1 pt.	2 pts.	3 pts.	4 pts.
No answer	Below expectation	Progressing	Meets criteria	Exemplary
No answer is written or element is missing	Answer shows serious misunderstanding and/or numerous errors	A few elements are present and correct, but little understanding is demonstrated	Majority of elements are present and correct; basic understanding is evident	All elements are present and correct; answer construction shows deep understanding

II. Analytic Rubric: An Analytical Rubric for Responses to the Challenge Statement, "Plants Get Their Food from the Soil"

Adapted from Table 3 in Allen and Tanner 2006

Criterion: Demonstrates an understanding that . . .	2 points	1 point	0 points
1. Food can be thought of as carbon-rich molecules including sugars and starches.	Defines food as sugars, carbon skeletons, or starches or glucose.	Attempts to define food and examples of food, but does not include sugars, carbon skeletons, or starches. Must go beyond use of word "food."	Does not address what could be meant by food or only talks about plants "eating" or absorbing dirt.
2. Food is a source of energy for living things.	Describes food as an energy source and discusses how living things use food.	Discusses how living things may use food, but does not associate food with energy.	Does not address the role of food.
3. Photosynthesis is a specific process that converts water and carbon dioxide into sugars.	Discusses photosynthesis in detail, including a description of the reactants (water and carbon dioxide), their conversion with energy from sunlight to form glucose/sugars, and the production of oxygen.	Partially discusses process of photosynthesis and may mention a subset of the reactants and products, but does not demonstrate understanding of photosynthesis as a process.	Does not address the process of photosynthesis. May say that plants need water and sunlight.
4. The purpose of photosynthesis is the production of food by plants.	Discusses the purpose of photosynthesis as the making of food and/or sugar and/or glucose by plants.	Associates photosynthesis with plants, but does not discuss photosynthesis as the making of food and/or sugar and/or glucose and/or starch.	Does not address the purpose of photosynthesis.
5. Soil may provide things other than food that plants need.	Discusses at least two appropriate roles for soil for some plants. Possible roles include the importance of minerals (N, P, K), vitamins, water, and structural support from the soil.	Discusses at least one appropriate role for soil. Possible roles include the importance of minerals (N, P, K), vitamins, water, and structural support from the soil.	Does not address an appropriate role for soil. The use of the word "nutrient" without further elaboration is insufficient for credit.

Appendix B: Resources for Creating and Using Rubrics

Association of American Colleges and Universities. VALUE: Valid Assessment of Learning in Undergraduate Education. Rubrics available online as PDFs: http://www.aacu.org/value/metarubrics.cfm

Carnegie Mellon. Enhanced Education Center: http://www.cmu.edu/teaching/designteach/teach/rubrics.html

iRubric (search for rubric to grade a scientific research paper): https://www.rcampus.com/indexrubric.cfm

MERLOT. Pedagogy Portal: http://pedagogy.merlot.org/RubricsandGrading.html

Michigan State University. Office of Faculty & Organizational Development: http://fod.msu.edu/oir/Assessment/rubrics.asp

Rubistar. Free online tool to create rubrics: http://rubistar.4teachers.org/

Scoring Rubrics. Field-tested Learning Assessment Guide (FLAG): http://www.wcer.wisc.edu/archive/cl1/flag/cat/catframe.htm

University of Minnesota. Center for Advanced Research on Language Acquisition (CARLA). Virtual Assessment Center. Creating Rubrics: http://www.carla.umn.edu/assessment/vac/Evaluation/p_7.html

Appendix C: Information on Calibrated Peer Review (CPR)™ and Immediate Feedback Assessment Technique (IF-AT)

More information on Calibrated Peer Review (CPR)™ can be found at the Project Kaleidoscope web site: http://www.pkal.org/documents/Vol4CalibratedPeerReview.cfm

More information on the Immediate Feedback Assessment Technique (IF-AT), including how to obtain the answer forms, can be found at: http://www.epsteineducation.com/home/about/default.aspx

4

Formative Assessment

Assessment *for* Learning

If they sit and listen, what they learn to do is to sit and listen.

—Randall Phillis

While both summative and formative assessment drive learning, they do so in different ways (Wiliam and Black 1996). As discussed in Chapter 3, summative assessment typically occurs at a particular milestone in a course or curriculum to evaluate student learning. In contrast, formative assessment occurs *during* the learning process and provides continual feedback to both the student and instructor (Bloom *et al.* 1971; Wiliam and Black 1996; Black and Wiliam 1998; Black and Wiliam 2004). Like active learning, formative assessment has been shown to improve student learning in the sciences and other disciplines (see Black and Wiliam 1998; Yuretich *et al.* 2001; Black and Wiliam 2004; Knight and Wood 2005; Michael 2006; Freeman *et al.* 2007; Reeves 2007; Thompson and Wiliam 2007; Froyd 2008; Walker *et al.* 2008; Foster and Poppers 2009). The utility of formative assessment in large part is due to its iterative nature; the ongoing feedback provides students with repeated opportunities to monitor their learning and reflect on their roles as learners. This process of metacognition, or being aware of one's own learning (Bransford *et al.* 1999), affords students multiple opportunities to weigh their understanding of the material and check it against the class expectations. Clear expectations—when assessment aligns with the intended learning outcomes and the classroom instruction—become paramount for this agenda. What better way to engage students in learning than to empower them with clear expectations, opportunities to adjust their approaches, and plenty of feedback? This chapter is designed to help instructors use formative assessment to help students learn to think metacognitively and achieve the intended learning outcomes.

Formative Assessment and Learning

Formative assessment functions in multiple capacities to facilitate learning. First, it actively involves students in the completion of tasks that promote construction of new knowledge or practice of desired skills. While attempting to perform these tasks, students must engage their prior knowledge. If the tasks are aligned to other course elements, the level of performance on the tasks will provide feedback about both students' prior knowledge and their progress toward the desired learning outcomes (Wiliam and Black 1996). This feedback can be used for important midstream adjustments. For example, if a probing question reveals that a significant portion of the class is having trouble with a concept, the instructor can use that information to address the issue before moving on to a new topic; the students, in turn, can use the feedback to identify problem areas and refocus their study efforts. Therefore, effective formative assessment requires students to ultimately construct new understanding using previous knowledge. Handelsman *et al.* (2007) coined the term "Engaugement" to capture the inextricable link between formative assessment, which *gauges* learning, and active learning, which *engages* students in learning (Figure 4.1). The following sections will elaborate on these elements of formative assessment.

Formative Assessment Engages Students in Active Learning

Formative assessment provides students with opportunities for "deliberate practice," or purposeful activities that promote improvement of a skill, which are necessary for the transition from novice to expert (Ericsson *et al.* 1993). Neither a music student learning to play chords nor a science major learning to design experiments can do so passively. Unfortunately, while college science faculty value the acquisition of scientific process skills by their students, few report using class time to teach these skills (Coil *et al.* 2010). Formative assessment activities

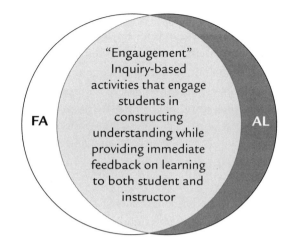

Figure 4.1. Engaugement. "Engaugement" is the term coined to describe the inter-section between formative assessment (FA), which provides ways to *gauge* student learning, and active learning (AL), which *engages* students in constructing new knowledge (Handelsman *et al.* 2007).

that require students to grapple with difficult concepts and think creatively pro-vide the type of deliberate practice necessary to acquire deep conceptual under-standing and acquire higher-level cognitive skills. Likewise, assessment activities that involve students in the processes of science help science majors develop the skills necessary to succeed in their disciplines.

Formative Assessment Engages Students' Prior Knowledge

Students bring their previous knowledge and experiences to bear when attempt-ing to answer questions or solve problems. John Dewey was among the first to challenge the idea of students as "empty vessels" waiting to be filled with knowl-edge (Dewey 1966). His view of learning led to the modern educational theory of constructivism, which posits that students learn by integrating new informa-tion or experiences with what they already know or think—prior knowledge or preconceptions (Ausubel 2000). Activities that engage prior knowledge help stu-dents reinforce correct preconceptions and confront and modify incorrect ones, thereby creating a new understanding (Bransford *et al.* 1999). While tasks that involve students physically or emotionally can enhance students' motivation or buy-in, these formative assessment activities must also engage students mentally to have an impact on learning. Engaging previous knowledge and confronting misconceptions are essential for learning. Documentaries like *A Private Universe* and *Minds of Our Own* (Schneps *et al.* 1989, 1997) highlight the difficulties students face in overcoming misconceptions and underscore how vanishingly small the chance is that they will do so as passive listeners in a lecture course.

Formative Assessment Provides Timely Feedback

Engaging students is only half the challenge. Timely feedback is necessary for students to gauge their progress and alter their study habits accordingly (Ramaprasad 1983; Black and Wiliam 1998). Research by Black and Wiliam (1998) determined that "ongoing assessment plays a key role—possibly the most important role—in shaping classroom standards and increasing learning gains." A crucial component of formative assessment is that it provides regular, ongoing feedback that allows students to reinforce accurate knowledge and modify misconceptions (Butler *et al.* 2007). Imagine how much less effective piano practice would be if students could not hear their music. Unable to detect the wrong notes, students would continue to make the same mistakes. Engaging in learning activities without feedback is like practicing on an electronic keyboard with the sound turned off.

Aligned Formative Assessment Facilitates Achievement of Intended Learning Outcomes

Finally, formative assessment activities are most effective when they align with the cognitive levels of both the intended learning outcomes and the summative assessments. Activities that do not help students achieve the desired outcomes are no more effective than passive coverage of content (Wiggins and McTighe 1998). Alignment is what makes the practice afforded by formative assessment deliberate. Just as musicians and athletes spend hours training, science students need time to practice thinking like scientists. In the classroom, this process can be accelerated with activities designed to help students develop skills that represent scientific ways of thinking.

Formative Assessment in Practice

A simple mantra for instructors attempting to engage students with formative assessment is "Ask, don't tell." Effective strategies employ questioning to engage students. Questions can capture students' interest and emphasize the significance of what they are learning. In contrast to simple presentation of information by the instructor, questions enhance learning by requiring students to recall information from memory, which promotes long-term retention. Frequent testing has been found to be more effective than repeated study or rereading for promoting long-term retention (Karpicke and Roediger 2007, 2008; Karpicke and Blunt 2011). Chapter 6 will further address this phenomenon, known as the testing effect (Glover 1989; Roediger and Karpicke 2006a, 2006b). Effective questions also provide students with opportunities to elaborate on answers, confront problems, and synthesize information, thereby building a deeper understanding of the material (Webb 1989; Chi *et al.* 1994; Coleman *et al.* 1997; Smith *et al.* 2009; Tanner 2009). Well-designed quiz questions, case studies, and group problem-solving activities are just a few examples of formative assessment activities that require students to retrieve previously learned information and employ critical thinking to answer questions.

Formative assessment can be implemented piecemeal, on a small scale, or as a large-scale reconfiguration of course structure. In small-scale implementation, formative assessment activities are incorporated gradually using any of the techniques described in Table 4.1. Another resource, *Classroom Assessment Techniques* (Angelo and Cross 1993), provides an extensive overview of assessment techniques, including but not limited to the common examples addressed here. In contrast, large-scale adoption of in-class formative assessment can involve extensive transformation of a traditionally taught course and thus can require a greater time commitment. Both small- and large-scale implementation have been shown to enhance student learning; the choice of implementation techniques will depend on their compatibility with intended learning outcomes, on instructor and student preferences, and on the availability of technology and other resources.

The majority of formative assessment techniques addressed in this chapter involve collaborative learning, broadly defined as students working in small groups to achieve a common goal (Haring-Smith 1993). The benefits of both collaborative learning and cooperative learning (which is usually more structured than collaborative learning) in the sciences have been well documented

(Johnson *et al.* 1990; Smith *et al.* 1991; Springer *et al.* 1999; Udovic *et al.* 2002; Beichner and Saul 2003; Ebert-May *et al.* 2003; Michael 2006; Wood 2009; Osborne 2010).[1] In addition to improving learning, collaborative activities allow students to practice skills necessary for working effectively in groups, which is considered essential for students to succeed in the twenty-first-century workplace (Partnership for 21st Century Skills 2009).

Small-Scale Formative Assessment

Incorporating formative assessment activities into a class takes time; instructors will not be able to add assessment activities and still lecture on all of the content previously covered. This necessitates a change in the way that time is used both in and out of class. As information has become easier for students to access outside of class, the instructor's traditional role of transmitting information by lecturing has become less important. When class time is used primarily for assessment activities that provide practice with difficult concepts and higher-order cognitive (HOC) skills, students must acquire lower-level information, facts, and terminology through pre-class activities; post-class assessment can offer additional practice and retrieval opportunities (Glover 1989; Roediger and Karpicke 2006a, 2006b). Class time is thereby freed for learning activities, without omitting necessary content. Small-scale formative assessment exercises can be added gradually and in a multitude of combinations to any course undergoing transformation into a learning-centered environment. Table 4.1 summarizes some common formative assessment strategies that can be used to facilitate this transformation (see the Appendix for further references).

Small-scale formative assessment activities can be used to support a variety of learning outcomes. For example, quizzes, immediate feedback assessment techniques (IF-AT), or reading assessments given prior to or at the beginning of class encourage students to become familiar with terminology and fact-based material required for HOC-level activities encountered during class. Given their

---

[1] While the terms "collaborative learning" and "cooperative learning" are often used interchangeably, collaborative learning is a less structured form of cooperative learning. Cooperative learning is characterized by activities that require cooperation and interdependence, where students take on prescribed roles with an emphasis on group dynamics and academic goals (Kagan 1994). Students involved in cooperative learning activities are assessed as individuals, while students involved in collaborative activities are assessed as a group (Johnson *et al.* 1998; Michael 2006).

Table 4.1. Examples of Small-Scale Formative Assessments

Adapted from Tables 3.1 and 3.2 of the *Vision and Change* report (AAAS 2011) and Table 3.1 in *Scientific Teaching* (Handelsman *et al.* 2007)

	Example	Intended Learning Outcome
Brainstorming	Students list everything they know on a given topic or in response to an open-ended question and then share responses with the class.	Students will retrieve prior knowledge and experiences related to a specific topic or query.
Case Studies	Students work through scenarios that require them to apply the science they are learning to social, ethical, political, or research-related dilemmas.	Students will evaluate a real-life problem and apply their knowledge and understanding to solve it.
Concept Maps	Students create a graphical illustration, consisting of arrows connecting boxes or circles, that represents relationships (arrows) between concepts (boxes/circles).	Students will synthesize their understanding of complex or abstract processes/systems.
Diagnostic Question Cluster (DQC)	Groups of questions intended to make student reasoning apparent and uncover misconceptions. Can be administered as clicker, post-quiz, or exam questions.	Students will apply scientific reasoning to solve problems.
Drawing Pictures or Diagrams	Students individually or in groups draw a picture or diagram representing a process in science.	Students will synthesize their knowledge and understanding of a system or process.
Immediate Feedback Assessment Technique (IF-AT)	Answer-until-correct interactive testing system where students get feedback as they answer. Intended to correct wrong answers immediately.	Students will retrieve and apply prior knowledge to answer questions and will employ feedback to correct misconceptions.
Immediate Response Systems ("Clickers")	Class polling systems that allow students to respond anonymously to questions, get immediate feedback on answers, and view class results.	Students will retrieve information to answer questions, apply understanding to solve problems, and employ feedback to correct misunderstandings.
Just-in-Time Teaching (JiTT)	Instructor uses student performance on pre-class web assignments to modify class instruction based students' prior knowledge.	Students will retrieve and apply prior knowledge to answer questions.
One-Minute Papers	Students write for one minute on a topic or in response to an open-ended question.	Students will synthesize their knowledge and understanding of a topic.
Quizzes	Frequent quizzing either at the beginning of class or online prior to or after class.	Students will retrieve and apply prior knowledge to answer questions.
Reading Assessments	Students work in groups to design an activity that will assess comprehension of a reading assignment. The group leads the activity in class.	Students will apply knowledge gained from reading assignments to complete a task.
Statement Correction	Students identify the error in a statement and discuss possible corrections with a partner or group.	Students will evaluate a statement and apply their understanding to correct mistakes.
Strip Sequence	Students work with a partner or group to place the steps of a process in the appropriate logical order.	Students will evaluate a process and apply their understanding to recreate the logical steps of the process.
Think-Pair-Share	Students contemplate a problem or question individually, discuss answers or solutions with a partner or group, and share agreed-upon answers with the whole class.	Students elaborate on and defend their answer to a question or their solution for a problem.

busy schedules, this strategic use of formative assessment motivates students to prepare for class and thus reap greater benefits from in-class activities. Just-in-Time Teaching (JiTT) is another useful method to encourage class preparation while providing insight into students' prior knowledge. Students perform a pre-class exercise, typically in an online format that is quickly assessed. The instructor gathers responses prior to class and uses them to guide decisions about what materials and activities will occur in class (Novak *et al.* 1999). Instructors can decide to skip material that students already know well or increase time on a topic they do not.

Once students arrive in class, questions posed using brainstorming, immediate response systems ("clickers"), or case studies can promote interest, reveal preconceptions, and target misconceptions. For example, brainstorming is one of the easiest ways to find out what students already know when introducing a new topic. The instructor poses a relatively simple, open-ended question about a topic, allows students to discuss it with a partner or group for a minute, and then invites groups to share with the whole class. As these are typically broad questions posed at the beginning of a topic, there is less pressure to provide a "right" answer and students are more likely to respond even in large classrooms. Modifying any question into a think-pair-share activity, where students discuss answers with their neighbors or group first, can increase the likelihood that students will feel confident enough to share their answers with the whole class (Allen and Tanner 2002).

Immediate response systems—individual polling systems that allow students to answer questions in real time using small remote responders—are effective tools to promote student participation and learning, especially in large classes (Brewer 2004; Caldwell 2007; Preszler *et al.* 2007; Smith *et al.* 2009; Levesque 2011). Typically in multiple-choice format, these "clicker" questions are posed using presentation software, and student responses are collected and displayed as a histogram. While instructors can track individual student responses in order to assign grades, from the students' perspective the responses are anonymous. This anonymity is important in increasing response rates above traditional polling methods such as hand-raising (Wood 2004). Colored index cards to represent the letters, or fingers to represent the numbers corresponding to different answer choices, are low-tech alternatives that preserve the ability to ask multiple-choice questions (MCQs) but not the anonymity of clickers.

As mentioned above, clicker questions can be used to elicit prior knowledge or confront misconceptions when introducing a topic in class. "Concept inventory" questions, another method to address misconceptions in a discipline, provide useful models for writing effective clicker questions of this type (see Chapter 7 for more information about concept inventories). In addition to revealing prior knowledge, questions can be written to elicit higher-order thinking by students (see Chapter 5 for detailed guidelines on writing HOC MCQs). Diagnostic question clusters (DQCs), question sets written with the explicit purpose of eliciting scientific reasoning by students, are particularly useful for this purpose. The distractors for these questions are chosen to illuminate student misconceptions (D'Avanzo 2008). For example, a question about energy flow through systems will contain distractors that reveal whether students harbor misconceptions about the conservation of matter or the conservation of energy, thereby providing insight into why students miss a particular question (Hartley *et al.* 2011; Maskiewicz *et al.* 2012). While clicker questions can be used to assess lower-order cognitive (LOC) level questions, the most effective types of clicker questions promote student elaboration, problem-solving, and peer teaching (Preszler *et al.* 2007; Smith *et al.* 2009; Levesque 2011). For example, when students work together to explain why answers are correct or incorrect, they practice critical thinking skills such as analysis and evaluation. Furthermore, questions that require students to analyze data or make predictions foster scientific ways of thinking. Allowing students to vote, discuss, and revote before showing the right answer is an effective way to promote explanation, peer teaching, and learning (Mazur 1997; Smith *et al.* 2009).

Like brainstorming and clicker questions, case studies are effective tools for engaging students with a new topic. Because case studies usually employ real-world scenarios, they provide an attention-grabbing context within which to embed a variety of formative assessment activities, including clicker questions (Smith *et al.* 2005; Herreid 2006; Wolter *et al.* 2011). For example, case studies can introduce a new topic or concept using a relevant problem that students cannot solve at first, while simultaneously providing the motivation for students to acquire the knowledge and skills to address the dilemma. After engaging in learning activities, students return to the case study's task to apply their knowledge and exercise their problem-solving skills. Case studies are therefore powerful instruments for promoting higher-order thinking—the type of thinking that is a hallmark of science—since students typically have to apply understanding, process information, and analyze data in order to solve problems, resolve dilemmas, or provide judgments on outcomes (Herreid 2006, 2007). Case studies have long been used for educational purposes by schools of law, business, and medicine. Even though more widespread adoption of this technique by science educators has been relatively recent, online databases like the National Center for Case Study Teaching in Science (http://sciencecases.lib.buffalo.edu/cs/) offer a plethora of cases that can be used by instructors teaching in the STEM disciplines at all educational levels, from middle school through professional development (see the Appendix for other online resources). Whether posed using clickers, think-pair-share activities, or case studies, questions that require students to formulate hypotheses, interpret graphs, and use data to draw conclusions facilitate acquisition of scientific process skills.

Several assessment activities—such as drawing exercises, concept maps, strip sequences, or one-minute papers—can prompt students to synthesize what they've learned into a big-picture understanding. Drawing exercises and one-minute papers allow students to express their understanding of a concept in their own images or words, which can be surprisingly effective at uncovering lingering misconceptions or gaps in understanding. Concept maps and strip sequences require students to apply what they have learned in order to connect concepts or put the steps of a process in logical order, and they give immediate feedback as to whether students understand the system or process (Novak and Gowin 1984; Jonassen *et al.* 1993). When administered at the end of a topic, these formative assessment strategies help instructors gauge the level of understanding prior to moving on to the next unit (McClure *et al.* 1999).

Finally, post-class assessment activities in the form of online quizzes offer students additional opportunities to practice retrieving material learned in class, which may be as valuable in driving learning as some in-class elaborative activities. Karpicke and Blunt (2011) found that activities requiring retrieval drive learning more effectively than activities involving concept maps (Karpicke and Blunt 2011). In addition, homework assignments provide additional practice with difficult concepts or skills that require more than a single exposure to master. These types of assignments provide important feedback for students who may have the misperception that they have mastered a skill merely because they were able to successfully complete the task during class under the guidance of their instructor.

Large-Scale Systems of Formative Assessment

The large-scale formative assessment frameworks addressed in this chapter (Table 4.2) typically involve transformation of the entire course and require a

Table 4.2. Large-Scale Systems of Formative Assessment

Adapted from Table 3.2 of the *Vision and Change* report (AAAS 2011)

	System	Learning Outcome
Model-Based Learning (MBL)	Students work with mental models to represent and comprehend complex phenomena.	Students will be able to render complex systems in a simplified, logical representation and develop a working understanding of processes or phenomena in a discipline.
Peer-Led Team Learning (PLTL)	Previous students serve as guides (peer leaders) and facilitate small-group discussions for current students.	Students will be able to apply their understanding of course concepts to solve problems.
Problem-Based Learning (PBL)	Students work through a cycle of evaluating a problem, collecting data, recommending solutions, and evaluating the process of coming up with a solution.	Students will develop formal problem-solving skills and a working understanding of course concepts.
Process-Oriented Guided Inquiry Learning (POGIL)	Students are assigned to specific roles within small groups so that all are actively engaged in the process of guided inquiry learning.	Students will develop disciplinary process skills to explore an observation or phenomenon in order to come to a new level of understanding.
Team-Based Learning (TBL)	Students work in permanent groups as high-performance teams to accomplish learning through structured tasks.	Students will be able to evaluate a problem in order to select and defend a solution using discipline-based knowledge.

larger up-front time commitment by the instructor. On the other hand, some of them may in fact be more attractive for active-learning novices because their implementation is highly structured. Process-oriented guided inquiry learning (POGIL) and team-based learning (TBL), for example, both prescribe in- and out-of-class work periods broken up into blocks or phases in which specific activities are scheduled to take place. When transitioning from teacher-centered to learner-centered classes, this level of structure may provide a feeling of direction and security for both the instructor and the student. Alternatively, problem-based learning (PBL), model-based learning (MBL) and case-based learning (case studies, Table 4.1) are approaches that can be implemented as small modules or as a course-wide framework. Instructors may start out using case studies, problems, or models in combination with other strategies and evolve toward a purely case-, model-, or problem-driven course. Finally, peer-led team learning (PLTL) differs from these other frameworks in its use of former students as learning assistants. These students act as mentors and facilitators of group work for current students in the course. While this arrangement can be beneficial for both peer-leaders and current students, it does take additional time and expertise to recruit and train peer leaders (Roscoe and Chi 2004; Gafney and Varma-Nelson 2008).

Teaching is both a personal and a public endeavor; there is no one-size-fits-all approach. When implementing formative assessment into a backward-designed class, the instructor should choose strategies that align with learning outcomes, suit personal preferences, and make the most of available space and technology. If the implementation is small-scale, using only one or two different techniques to start will allow sufficient time for the instructor to prepare new materials and for both the instructor and the students to become familiar with the new format. For example, using daily brainstorming to elicit prior knowledge and posing clicker questions to confront misconceptions and practice problem-solving could be a manageable approach to easing formative assessment into a previously traditional classroom. In addition, by being transparent about the rationale for the new teaching strategies, instructors afford students the opportunity to become partners in the learning process rather than passive consumers.

Conclusion

Effective formative assessment engages students in tasks that elicit prior knowledge, build new understanding, foster skill development, and provide valuable feedback on student progress toward intended learning outcomes. This feedback allows students to modify their behavior to better achieve the learning outcomes and allows instructors to make real-time modifications of instructional materials to address problems in understanding. Well-aligned formative assessment offers science students the practice and timely feedback necessary to master vital content knowledge, build conceptual understanding, and practice scientific process skills. In sum, formative assessment can be one of the most powerful tools in an instructor's toolbox of teaching methods.

Appendix

Table 4.1. Examples of Small-Scale Formative Assessments (with references)

Adapted from Tables 3.1 and 3.2 of the *Vision and Change* report (AAAS 2011) and Table 3.1 in *Scientific Teaching* (Handelsman *et al.* 2007)

	Example	Reference
Brainstorming	Students list everything they know on a given topic or in response to an open-ended question and then share responses with the class.	
Case Studies	Students work through scenarios that require them to apply the science they are learning to social, ethical, political, or research-related dilemmas.	Boehrer and Linsky 1990; Allen and Tanner 2003; Herreid 2006, 2007, 2011 http://sciencecases.lib.buffalo.edu/cs/ http://www.caseitproject.org/ http://www.bioquest.org/icbl/cases.php
Concept Maps	Students create a graphical illustration, consisting of arrows connecting boxes or circles, that represents relationships (arrows) between concepts (boxes/circles).	Novak and Gowin 1984; Novak 2005; Novak and Cañas 2006; Lim *et al.* 2009
Diagnostic Question Cluster (DQC)	Groups of questions intended to make student reasoning apparent and uncover misconceptions. Can be administered as clicker, post-quiz, or exam questions.	Maskiewicz *et al.* 2012
Drawing Pictures or Diagrams	Students individually or in groups draw a picture or diagram representing a process in science.	
Immediate Feedback Assessment Technique (IF-AT)	Answer-until-correct interactive testing system where students get feedback as they answer. Intended to correct wrong answers immediately.	Epstein *et al.* 2002; Cotner *et al.* 2008
Immediate Response Systems ("Clickers")	Class polling systems that allow students to respond anonymously to questions, get immediate feedback on answers, and view class results.	Brewer 2004; Guthrie and Carlin 2005; Hall *et al.* 2005; Caldwell 2007; Mayer *et al.* 2009 http://srri.umass.edu/topics/crs/bibliography
Just-in-Time Teaching (JiTT)	Instructor uses student performance on pre-class web assignments to modify class instruction based students' prior knowledge.	Novak *et al.* 1999
One-Minute Papers	Students write for one minute on a topic or in response to an open-ended question.	

	Example	Reference
Quizzes	Frequent quizzing either at the beginning of class or online prior to or after class.	Klionsky 2004; Karpicke and Roediger 2007, 2008; McDaniel *et al.* 2007
Reading Assessments	Students work in groups to design an activity that will assess comprehension of a reading assignment. The group leads the activity in class.	Handelsman *et al.* 2007
Statement Correction	Students identify the error in a statement and discuss possible corrections with a partner or group.	
Strip Sequence	Students work with a partner or group to place the steps of a process in the appropriate logical order.	
Think-Pair-Share	Students contemplate a problem or question individually, discuss answers or solutions with a partner or group, and share agreed-upon answers with the whole class.	Allen and Tanner 2002

Table 4.2. Large-Scale Systems of Formative Assessment (with references)

Adapted from Table 3.2 of the *Vision and Change* report (AAAS 2011)

	System	Reference
Model-Based Learning (MBL)	Students work with mental models to represent and comprehend complex phenomena.	Gilbert and Boulter 1998; Buckley 2000
Peer-Led Team Learning (PLTL)	Previous students serve as guides (peer leaders) and facilitate small-group discussions for current students.	Gosser *et al.* 2001; Roscoe and Chi 2004; Eberlein *et al.* 2008; Gafney and Varma-Nelson 2008 http://www.pltl.org/
Problem-Based Learning (PBL)	Students work through a cycle of evaluating a problem, collecting data, recommending solutions, and evaluating the process of coming up with a solution.	Allen 1997; Waterman and Stanley 1998; Allen and Tanner 2003; Eberlein *et al.* 2008 http://www.studygs.net/pbl.htm
Process-Oriented Guided Inquiry Learning (POGIL)	Students assigned to specific roles within small groups so that all are actively engaged in the process of guided inquiry learning.	Farrell *et al.* 1999; Moog and Spencer 2008; Schroeder and Greenbowe 2008; Moog *et al.* 2009; Brown 2010 http://www.pogil.org/
Team-Based Learning (TBL)	Students work in permanent groups as high-performance teams to accomplish learning through structured tasks.	Michaelsen *et al.* 2002, 2004; Barkley *et al.* 2005 http://www.teambasedlearning.org/

5

Assessing Higher-Order Cognitive Skills with Multiple-Choice Questions

> Most teachers waste their time by asking questions which are intended to discover what a pupil does not know, whereas the true art of questioning has for its purpose to discover what the pupil knows or is capable of knowing.
>
> —Albert Einstein

Multiple-choice exams are a fact of life for instructors teaching large courses. These instructors face the challenge of effectively assessing their students' critical thinking and problem-solving skills (higher-order cognitive skills, or HOC skills) while trying to keep the grading workload manageable. Using exams comprised exclusively of essay or short-answer questions is impractical. However, a well-designed multiple-choice exam with a reasonable number of HOC questions can evaluate higher-order thinking while minimizing grading time. Part of the time gained through faster grading is spent designing HOC multiple-choice questions (MCQs), which is difficult. However, in the long run, the time investment will pay off in the form of a test bank large enough to make reusing questions practical (Herskovic 1999). This chapter provides practical guidance for writing different types of MCQs that can evaluate HOC skills.

Multiple-Choice Exams

Multiple-choice exams have received criticism in education literature because these exams often assess mostly factual recall and not higher-order reasoning (Bloom 1956; Zheng *et al.* 2008; Momsen *et al.* 2010; Shepard 2010). Indeed, an analysis of nearly ten thousand assessment items from faculty teaching introductory biology showed that 93 percent of exam questions were at the lowest two Bloom levels (*remember* and *understand*), even though faculty stated that they wanted their students working at much higher levels (Momsen *et al.* 2010). It is not surprising that recall questions tend to dominate multiple-choice exams, as these questions are the easiest to write and most instructors have not had training in writing HOC multiple-choice questions. However, by applying some of the same techniques that professional exam writers use, faculty can increase the cognitive level of their MCQs and thereby better align exams with their intended learning outcomes.

Beyond reducing the workload associated with grading, there are other advantages of multiple-choice exams. Students can answer more MCQs than free-response questions in a given time period. The more questions a student answers, the more broadly the course content is sampled. Broader sampling increases the content validity of an exam, meaning the exam measures the content it was designed to measure (Engelhardt 2009). In addition, a greater number of exam questions allows instructors to better differentiate between students of varying abilities (ibid.). Lastly, in contrast to free-response questions, multiple-choice exams are necessarily scored objectively, since there is usually only one correct answer for each question.

Some of the disadvantages of multiple-choice exams, particularly in science, were described in Chapter 3. One such disadvantage is that MCQs provide students with choices from which they select an answer, leaving little room for demonstrating creative thinking. Also, MCQs do not readily capture students' achievement in scientific process skills, such as experimental design or written communication. As described in Chapter 2, lower-order cognitive (LOC) questions primarily test content recall, whereas HOC questions require students to apply, analyze, evaluate, or create information (Bloom 1956; Zoller 1993; Dickie 2003; Crowe *et al.* 2008). Many agree that MCQs cannot test at the *create* level of Bloom's Taxonomy, because this level requires students to generate something new from the information provided (Bloom 1956; Woodford and Bancroft 2005), but MCQs *can* test performance at all other Bloom levels. However, writing good HOC questions takes practice. Unlike free-response questions, MCQs are much more likely to cue the student to the correct answer because the answer is present (Palmer and Devitt 2007). Nevertheless, this pitfall can be avoided by writing questions skillfully. The following section offers some advice and tools for writing good MCQs across the spectrum of cognitive levels.

Multiple-Choice Exams in Practice

Prior to writing exam questions, many instructors find it useful to make an Exam Construction Guide that outlines the content covered by the exam and the general levels at which the students practiced with the content during the course (Table 5.1). The guide is an extension of the alignment table introduced in Chapter 2 (Table 2.2). It lists the percentages of course time devoted to each of the general topics covered, and subdivides these percentages to indicate the proportions that students spent working on either LOC or HOC skills. The table then serves as a guide for creating the appropriate number of exam questions at the corresponding cognitive levels. For example, the sample Exam Construction Guide shown in Table 5.1 indicates that most of the HOC questions for this exam should focus on general thermodynamics and membrane permeability. Therefore, the HOC questions for these content areas should be written first, since they will most strongly affect the overall length of the test.

Table 5.1. Example of an Exam Construction Guide

Topics	% Time	LOCs	HOCs
Proteins and enzymes	20%	60%	40%
General thermodynamics	25%	20%	80%
Membrane permeability	20%	35%	65%
Diffusion and osmosis	10%	50%	50%
Cellular Respiration	25%	75%	25%

Time and Cognitive Levels at which Students Practiced Content

Structure and Format of Multiple-Choice Questions

The structure of a standard MCQ is shown in Figure 5.1, but there are many other formats of MCQs that we will discuss in detail later in this chapter. The standard MCQ includes a "stem" that presents the reader with either a question or a sentence to be completed, followed by possible answers or phrases, which are referred to as options. Correct answers are called *keys*, while incorrect options are referred to as *distractors*.

Stem { What difference between RNA and DNA accounts for why RNA can act as a catalyst but DNA cannot?

a. RNA is single-stranded and DNA is double-stranded
b. RNA contains uracil whereas DNA does not
c. RNA is produced by transcription whereas DNA is produced by replication } **Distractors**
Key → d. RNA is made of ribose whereas DNA is made of deoxyribose

Figure 5.1. Components of a Standard Multiple-Choice Question at the LOC Level

Good distractors are based on common misconceptions students have about the content. While using two to five distractors is common, evidence suggests that three is optimal (Rodriguez 2005).

Standard MCQs can be written as either questions or partial sentences. Research has shown no real difference in overall effectiveness between these two formats, but for partial sentences, care must be taken with the grammatical structure of the options (Wilson and Coyle 1991; Woodford and Bancroft 2005). Stems that require students to complete a sentence must be written clearly so that distractors can't be easily eliminated because of grammatical disagreement. This is particularly true when choosing the articles "a" or "an," or writing distractor phrases beginning with a verb. Poorly written distractors can lead students to select the wrong option or can make wrong options so obvious that the question becomes meaningless to assess student learning.

The following example illustrates poorly written distractors; two options (a and c) can easily be eliminated because the article in the stem does not agree with the first word in the option. Making the article part of the distractor bypasses this pitfall.

This chapter contains an ________

a. step-wise guide for writing LOC MCQs
b. example of a standard multiple-choice question
c. table describing the percentage of overall questions an exam should have
d. opening quote about the use of active learning in the classroom

MCQs can take several forms of the standard structure (Table 5.2). The type of MCQ selected depends on the content and cognitive level being assessed as well as the time limit for the exam. Some MCQs take much longer to answer than others because they are inherently more difficult. For example, the questions may require students to analyze data or diagrams, or to perform calculations. In other cases, certain MCQ structures lend themselves to testing only LOC skills, such as matching and simple true/false formats that generally test only a student's ability to recall facts. (See Chapter 2 for more on determining the cognitive levels of questions.)

Writing Multiple-Choice Questions

Guidelines for writing effective MCQs can help instructors construct exams that are both rigorous and fair. The ability of MCQs to consistently discriminate between students with different academic abilities is referred to as *discriminatory power*. Both the discriminatory power and the overall reliability of multiple-choice exams can be enhanced by following the basic guidelines outlined in Table 5.3. Appendix A provides more in-depth explanations of different MCQ formats and how they relate to cognitive levels.

Prior to writing an HOC multiple-choice question, determine your approach and target an HOC level. There are three basic approaches to writing HOC multiple-choice questions: 1) writing from scratch (most time-consuming); 2) converting an HOC free-response question into a multiple-choice question (most time-efficient); and 3) converting an LOC multiple-choice question into an HOC multiple-choice question. The third method may be most effectively implemented by rewriting the LOC question as a free-response question and then converting it into an HOC multiple-choice question (see examples below). Once you have selected an approach, determine at which HOC level the question will be written. HOC multiple-choice questions can assess the following Bloom levels: *application*, *analysis*, and *evaluation* (see Chapter 2 for a full review of Bloom's Taxonomy). *Synthesis* generally cannot be assessed using multiple-choice questions because requiring a student to design an experiment or to create a graph or drawing to answer a question is not possible in a multiple-choice question format. Table 5.4 contains a description of the HOC Bloom levels that can be used to guide construction of HOC multiple-choice questions.

Table 5.2. Common Formats of Multiple-Choice Questions

(See Appendix A for examples)

Multiple-Choice Question Format	Advantages	Disadvantages
Standard A question or unfinished sentence requiring a forced selection from 4 options	Useful for measuring HOC skills if strong distractors are used	If not well-written, the structure can cue a student to the answer
Context-Dependent A scenario followed by a series of standard MCQs based on that scenario	Useful for measuring HOC skills if strong distractors are used	Takes relatively longer to answer
Two-Tiered An MCQ followed by either another MCQ or a free-response question requiring students to justify their reasoning for selecting their answer to the first question	Useful for illuminating student misconceptions	Takes relatively longer to answer Answer to one tier can cue the answer to the other tier
Complex (K-Type) A question followed by multiple possible answers, among which there may be more than one correct answer; the options can include groups of answers	Requires analytical skills of comparing and contrasting	Takes relatively longer to answer Tends to be misleading because there may be multiple correct answers
Matching Several words or phrases that are matched with one or more words or phrases	Takes relatively less time to answer	Typically not useful for measuring HOC skills
True/False One or more phrases followed by a true or false option; the phrase(s) may be associated with scenario, diagram, or data	Takes relatively less time to answer	True/false questions have increased errors associated with guessing
Multiple True/False One or more phrases followed by several true or false option(s); the phrase(s) may be associated with scenario, diagram, or data	Useful for measuring HOC skills	Takes relatively longer to answer
Alternative Choice A phrase, scenario, or other introduction followed by only two options, where one is a strong distractor	Useful for measuring HOC skills	Takes relatively longer to answer Less discriminating than standard format

Table 5.3. Steps in Writing Multiple-Choice Questions

Step	Task	Do	Avoid
1.	Determine the cognitive level of your question and select the appropriate MCQ format.	Analyze your Exam Construction Guide and confirm alignment between the question and the instruction.	Use of matching for HOC questions. Use of the complex MCQ format.
2.	Write the stem in a single, complete sentence if possible.	Be as concise as possible when writing the stem. Check for grammatical errors. If the stem requires the student to complete the sentence, confirm that the articles and verbs are in grammatical agreement with all options.	Use of the word "NOT" in the stem. Use of double negatives. Use of absolutes such as "always" or "never," as they can cue students to the correct answer.
3.	Write the key as *the* answer, not as the best answer.	If the stem requires the student to complete the sentence, place any related article "a" or "an" in front of key (and distractors) so that all options are grammatically correct.	Use of "none of the above", "all of the above," or "both C and D," as these can cue students to the correct answer or be misleading.
4.	Write three plausible distractors and a key. Arrange all options in a random manner.	Use strong distractors (those that are common misconceptions).	Use of weak distractors, as they cue students to the answer.
5.	Construct the test with all questions.	Check that the arrangement of correct answers for the entire test is random.	Placement of questions in an order that cues students to correctly answer a subsequent question.
6.	Review the exam and estimate the time it would take a student to answer each question.	Total up the percentage of LOC and HOC questions for each topic and compare totals to the Exam Construction Guide. Have a colleague review the questions.	Adding new questions without adjusting the time.

The quality of the distractors strongly influences the quality of the MCQ. This is particularly true for HOC MCQs. Plausible distractors based on misconceptions require students to engage their understanding and cognitive skills to answer the question, while implausible distractors can undermine a well-written stem. An effective strategy for generating good distractors is to collect student responses to a free-response question and use the common misconceptions displayed in their answers as the basis for the distractors for an MCQ on the same topic.

Table 5.4. HOC Bloom Levels for Multiple-Choice Questions

Bloom Level	Verbs Associated with HOC Multiple-Choice Questions	Student Actions for Solving HOC Multiple-Choice Questions
Evaluate	appraise, assess, conclude, criticize, critique, decide, defend, evaluate, judge, justify, prioritize, rank, rate, select, support, validate	Students judge the merit of information provided in order to select an option. The stem may contain information such as an experimental design or experimental data that the student must analyze; the associated options are critiques or statements in defense of the information. The student must evaluate all the information and select an option.
Analyze	analyze, break down, categorize, characterize, classify, compare, contrast, correlate, debate, deduce, diagram, differentiate, discriminate, distinguish, examine, infer, outline, question, rearrange, relate, separate, subdivide, test	Based on information given, students select an option that is an inference. Alternatively, they may have to discriminate between two options, or analyze a graph, data, or drawing to select an option.
Apply	act, administer, apply, calculate, change, chart, compute, demonstrate, determine, dramatize, employ, extend, illustrate, implement, inform, instruct, operate, practice, predict, prepare, produce, provide, role-play, show, sketch, solve, transfer, use, utilize	Students apply their knowledge in a novel way. This may require them to perform a calculation or use one or more different kinds of knowledge in order to select an option.

Converting Free-Response Questions into HOC MCQs

Now let's look at three examples of converting free-response questions into different formats of HOC MCQs. An in-depth explanation of the Bloom categorization and the caveats associated with certain formats accompany each example. The same considerations for converting free-response questions into MCQs should be used for writing MCQs from scratch. Both start with a general framework—the content measured, the scientific process skill required, and the format of the MCQ. Formats that are the most suitable for HOC MCQs are standard, context-dependent, two-tiered, true/false, and alternative choice. However, it should be noted that it is challenging to write a simple HOC true/false question unless it is a multiple true/false question for which a student must answer all statements correctly.

Example 1

Structure of MCQ	Standard MCQ
Cognitive Level	LOC/HOC; Apply
Discipline	General Chemistry (McKinstry, L., personal communication)
Free-response	Calculate the molar concentration of a 50.0 mL aqueous solution containing 25.0 g of glucose ($C_6H_{12}O_6$).
MCQ (answer in bold)	What is the molar concentration of a 50.0 mL aqueous solution containing 25.0 g of glucose ($C_6H_{12}O_6$)?

    a. 0.500 M
    b. 2.78 M
    c. 5.00×10^2 M
    d. 2.78×10^{-3} M
    e. 0.360 M

Explanation

This HOC question requires application of content about molar mass, moles, and molarity (Molarity = mol/L). Students must first determine the molar mass of glucose using the periodic table. They then convert the grams glucose into moles glucose using the molar mass, and convert milliliters into liters. Taking significant figures into consideration, they then use these values in the equation to solve for molarity. The student then selects the answer from the list of options. For these reasons, this question is at the *application* level of Bloom's Taxonomy.

The distractors for this question are plausible because if a student uses the number of grams given instead of converting it to moles, their answer would be close to option c. If they don't convert to liters and divide by grams, their answer is option a. Option e is a value close to one obtained if the dimensional analysis (unit conversion) is done incorrectly. Option d looks superficially similar to the correct answer. Although these are all plausible distractors, this question could be reduced to three options, a, b, and e, because those distractors are the most plausible given common mistakes with these types of calculations.

Example 2

Structure of MCQ	Context-Dependent MCQ
Cognitive Level	HOC; Analyze
Discipline	Interdisciplinary Science: Geology and Biology
Free-response	Describe how geologists use the principles of cross-cutting and superposition to determine the relative age of strata. How could this information be used as evidence that species change through time?

**MCQ
(answer in bold)**

1. Using the principles of cross-cutting and superposition, place the sequence of depositional or structural events depicted in the figure from oldest to youngest.

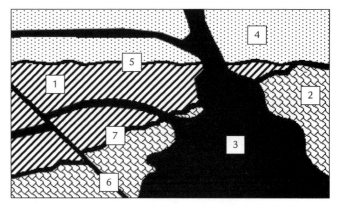

a. 2, 1, 7, 5, 4, 6, 3
b. 7, 2, 1, 6, 3, 5, 4
c. 2, 7, 1, 5, 4, 3, 6

2. If transitional fossils were found in layers 1, 2, and 4, which of the following is true of the fossils found in layer 1?
   a. fossils in layer 1 represent organisms from which the species in layer 4 evolved
   b. fossils in layer 1 represent organisms from which the species in layer 2 evolved
   c. fossils in layer 1 are not related to the species in layers 2 and 4

Explanation

This HOC question requires application of content, analysis of the diagram, and inference, and is at the *analysis* level of Bloom's Taxonomy. In the first part, students must apply the principles of superposition and cross-cutting to determine the relative dates of the strata. Strata deposited on top of other strata are younger; strata that cross-cut other strata are also younger. This question has two strong distractors because strata 2 and 7 are the oldest layers; it may appear that layer 7 cross-cuts layer 2, so some students might select this option. The students need only to compare three strata (1, 2, and 7) to make a selection.

Although the second question also requires students to determine the relative dates of strata, the three strata selected are depositional and easier to date. Thus, the task is not as difficult as in question 1. However, question 2 requires students to understand that transitional fossils show common ancestry and to infer the ancestral relationships by comparing the strata in which they are found. This question is at the *analysis* level of Bloom's Taxonomy.

Example 3

Structure of MCQ	True/False, Two-Tier Question
Cognitive Level	HOC; Apply or Analyze
Discipline	General Biology
Free-response	Explain how 2,4-dinitrophenol, an ionophore that transports protons across the inner mitochondrial membrane to the matrix, affects ATP synthesis.
MCQ (answer in bold)	1. The drug 2,4-dinitrophenol is an ionophore that transports protons across the inner mitochondrial membrane to the matrix. This drug would increase ATP synthesis. True or **False**? 2. Your selection is most closely associated with which of the following explanations? a. the drug directly increases the electrochemical gradient **b. the drug directly decreases the electrochemical gradient** c. the drug directly impacts ATP synthase and the reaction that makes ATP
Explanation	This HOC question requires application of content and inference of how 2,4-dinitrophenol impacts cellular respiration. To correctly answer the true/false question, students must know that the electron transport chain creates an electrochemical gradient, or proton force, by pumping protons from the matrix to the inner membrane space of the mitochondria. This gradient is then used to drive the reaction catalyzed by ATP synthase that makes ATP. Asking this question as a two-tiered question helps reduce the error associated with guessing the answer to the first question. Distractor "a" in question 2 is based on whether the student knows that the proton force is an electrochemical gradient and also the direction in which protons are pumped—from the matrix to the inner membrane space. Distractor "c" is based on a common misconception that the electron transport chain is directly connected to the production of ATP by ATP synthase. This drug would indirectly impact ATP synthase in catalyzing the reaction that makes ATP. This question is at the *analysis* level of Bloom's Taxonomy.

Conclusion

Faculty who have large classes often choose to assess their students with multiple-choice exams because MCQs can be graded significantly more quickly and easily than free-response questions. While MCQs have their drawbacks, well-written MCQs provide an effective way to assess student understanding at both LOC and HOC levels. The format of an MCQ often determines the cognitive level at which the student is assessed. Exam Construction Guides help faculty align assessment with instruction by providing an overview of the proportions of LOC- and HOC-level questions for each topic. Not surprisingly, MCQs that assess at HOC levels are more challenging and time-consuming to construct than LOC MCQs. However, by following a few simple guidelines, faculty can develop a reliable approach to creating effective HOC MCQs. Generating an HOC MCQ database is a worthwhile time investment, as such a database aids in the construction of summative assessment that challenges students to solve problems and use critical thinking skills. Moreover, it can serve as an excellent source of challenging in-class clicker questions.

Appendix A: Examples of Different Multiple-Choice Question Structures

(Described in Table 5.2)

Example 1: Context-Dependent

Description	Context-dependent MCQs are prefaced with a scenario, data table, graph, or similar introduction, followed by an item or series of items having two or more options. Answers to subsequent questions do not depend on the answer to previous questions, but all are related to the information found in the opening description of the scenario.
Cognitive Level	HOC and LOC
Discipline	Astronomy (Green 2002)
Question (answer in bold)	A star with a continuous spectrum shines through a cool interstellar cloud composed primarily of hydrogen. The cloud is falling inward toward the star (and away from Earth). An earthbound observer views the twinkling star.

Which best describes the spectrum seen by an earthbound observer?

a. blueshifted hydrogen emission lines
b. blueshifted hydrogen absorption lines
c. redshifted hydrogen emission lines
d. redshifted hydrogen absorption lines
e. a redshifted hydrogen continuum

The reason the observer views the star as twinkling is because of motion

_____ .

a. on the star's surface
b. of Earth
c. of the solar system
d. of gas in Earth's atmosphere

Advantages/ Disadvantages	Context-dependent questions are useful for measuring HOC skills because they usually require the student to analyze and problem-solve. These kinds of questions are very effective in science because students often have to make inferences from experimental data or diagrams, or evaluate a scenario before arriving at an answer. Multiple questions based on one graph or data set also provide an opportunity to measure several aspects of a student's understanding about a given topic. The disadvantages to these kinds of questions are that they usually take longer to answer, and the answer to one item can influence how the student answers the next item.

Example 2: Two-Tiered

Description

Two-tiered MCQs are similar to context-dependent MCQs in that they include more than one question. However, the two-tiered MCQs differ because the first MCQ requires a selection, and the second MCQ probes why the respondent selected the answer to the first question.

Cognitive Level

HOC

Discipline

Biology

Question (answer in bold)

You have tried different combinations of fertilizers to find the ideal mix to use on your favorite heirloom tomatoes. You set up an experiment to test your old and new fertilizer mixture. For each trial, you plant 30 of the same variety of heirloom tomato plants in the same soil under the following conditions:

Variable	trial 1	trial 2	trial 3	trial 4
fertilizer mix	old	new	old	new
sunlight (hours/day)	8	12	8	8
water (mL/day)	500	300	300	300
# tomatoes produced	80	140	100	100

1. Consider the following hypothesis: "The new fertilizer mixture affects the number of tomatoes a plant can produce." If the fertilizer experiment can be used to test the hypothesis, which trials should be compared?
   a. trials 1 and 2
   b. trials 2 and 3
   c. trials 3 and 4
   d. all trials
   e. none of the trials

2. Your selection is most closely associated with which of the following explanations?
   a. One should compare trials that show the new fertilizer yields many more tomatoes than the old fertilizer.
   b. One should compare trials that show a difference in the number of tomatoes yielded.
   c. One should compare all trials because conclusions can only be drawn from experiments that have been repeated numerous times.
   d. One should only compare trials that isolate the variable tested.
   e. One should only compare trials that support the hypothesis tested.

Advantages/ Disadvantages	The advantage to using two-tiered questions is that students who use test-taking strategies to correctly answer the first-tier question must still justify their answer in the second tier. Also, the second-tier options are usually based on common misconceptions; faculty can use this strategy to determine which misconceptions their students have. However, the disadvantage of having the second-tier MCQ is that it may cue a student to the correct answer in the first tier (Palmer and Devitt 2007). For this reason, many two-tier questions use an MCQ for the first tier and a free-response for the second tier (Treagust 1988).

Example 3: Complex (K-Type)

Description	Complex MCQs, also commonly referred to as K-Type questions, have several grouped options from which one must choose the correct answer or a group of answers from several options.
Cognitive Level	HOC
Discipline	Physics/Chemistry (ETS 2012)
Question (answer in bold)	Which of the following functions could represent the radial wave function for an electron in an atom? (r is the distance of the electron from the nucleus; A and b are constants.)

     I. $A e^{-br}$

     II. $A \sin(br)$

     III. A/r

a. I only
b. II only
c. I and II only
d. I and III only
e. I, II, and III

Advantages/ Disadvantages	Most agree that these kinds of questions are very confusing and often misleading for students (Albanese 1993; Haladyna *et al.* 2002). These questions also take up more space and require more time for students to answer. Instructors should avoid using them if at all possible. An alternative to this kind of question is the multiple true/false question (see Example 6).

Example 6: Multiple True/False

Description   Multiple true/false questions provide one stem followed by several statements about the stem that one must identify as true or false.

Cognitive Level   HOC

Discipline   Biology

Question (answer in bold)   A cartoon of an electron micrograph is shown below. A piece of chromosomal DNA containing a single gene is illustrated as a thick line. Thin lines radiating from the DNA are mRNA molecules in the process of being transcribed. Ribosomes are attached to the mRNA. *Newly synthesized peptides are not shown.*

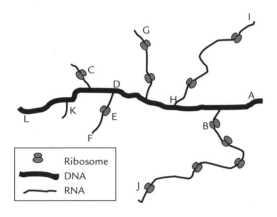

For the following statements, respond with true (t) or false (f).

The drawing above shows:

__**f**__   The sequence of bases in the mRNA labeled K is complementary to the sequence of bases in the mRNA labeled F.

__**t**__   The promoter for this gene is closer to L than to A.

__**t**__   Ribosome J has made fewer peptide bonds than ribosome B.

__**f**__   In the RNA transcript labeled H-I, the 5′ end of the RNA is at H.

__**t**__   The RNA polymerase located at position D (directly above E) is moving toward the right.

__**f**__   The cell from which this chromosomal DNA was isolated is eukaryotic.

__**t**__   The distance between ribosome E and end F is increasing.

__**f**__   Ribosome C has just synthesized the carboxy terminus of the protein.

Advantages/ Disadvantages   The example above shows how multiple true/false questions can be written to measure HOC skills: students must analyze a diagram and apply their knowledge of RNA transcription and protein translation to answer the questions. These kinds of questions are associated with increased difficulty (Frisbie 1990). However, the disadvantage of these kinds of questions is that they take longer to answer than other types of MCQs and they often have lower reliability (ibid.).

Example 4: Matching

Description   Matching questions provide several stems ar

Cognitive Level   LOC

Discipline   Immunology

Question
(answer in bold)

a. neutrophils	____D____
b. cytokines	____C____
c. platelets	____E____
d. chemokines	____A____
e. antigens	____B____

Advantages/
Disadvantages   Although little research has been done to de
questions are at measuring students' unders
questions are most likely to probe students'
ing requires identification based on memori;
essarily the cognitive level, of the question ca
items to match, but there is little research to
kinds of questions (Haladyna *et al.* 2002).

Example 5: Simple True/False

Description   Standard true/false questions provide a stat
true or false. These questions are also preser
requiring a student to make a choice betwee

Cognitive Level   LOC

Discipline   Geology (Coughenour, C., personal commu

Question   A Richter magnitude 6 earthquake has ten ti
(answer in bold)   magnitude 5 earthquake.

                                    True (t) or false (f)? _

Advantages/
Disadvantages   True/false questions typically measure recall
be used to assess a lot of factual recall of co
tions have increased errors associated with ;
HOC skills.

Example 7: Alternative-Option

Description	This is similar to a standard MCQ, but because it only has two options, it is called an alternative-option question. One of the options is a distractor.
Cognitive Level	HOC
Discipline	Chemistry

Question (answer in bold)

Based on the information provided below, how would you describe this reaction?

$$C_2H_4(g) + H_2O(l) \rightarrow C_2H_5OH(l)$$

$$\Delta G^{\circ}_{f}(C_2H_5OH(l)) = -175 \text{ kJ/mol}$$
$$\Delta G^{\circ}_{f}(C_2H_4(g)) \quad = 68 \text{ kJ/mol}$$
$$\Delta G^{\circ}_{f}(H_2O(l)) \quad = -237 \text{ kJ/mol}$$

a. The reaction is spontaneous.
b. The reaction is at equilibrium

Advantages/ Disadvantages

A two-option question can be just as effective as a standard MCQ with three or four options because many students easily eliminate options and make their choices based on two options anyway. With only two options, which are shorter than standard MCQs, an advantage of alternative-option questions is that more of them can be included on a test, thereby increasing the content that can be assessed (Haladyna *et al.* 2002).

CHAPTER

6

Preparing Students for Assessment

The key to surviving in an ever more rapidly changing and complex world is learning how to learn.

—Robert Bjork

Every instructor who has taught introductory college science has dealt with students who struggle with how to study for exams: they attend classes, highlight text in their books, make flash cards, and rewrite notes. Unfortunately, these activities increase the students' superficial exposure to the material without requiring them to engage with the concepts in any depth or to practice critical thinking skills. Such activities lead to what cognitive scientists refer to as the "illusion of knowing" (Jacoby *et al*. 1998; deWinstanley and Bjork 2002; Koriat and Bjork 2005). With each review of the material, students build a greater sense of familiarity. The words are no longer foreign, the figures seem less complicated, and the topic begins to feel manageable. As a result, students are lulled into a false sense of mastery of the material. While these strategies may suffice for courses that rely on memorization and recall, they are woefully inadequate for mastering higher-order cognitive (HOC) skills. Deep and meaningful learning requires that students actively engage with new material and integrate it with what they already know.

Students' approaches to learning fall into two broad categories: superficial or deep (Marton and Saljo 1976). Students taking a superficial approach see studying merely as a way to earn a grade, so they put in the minimum time needed to pass the class. These students spend little time integrating their knowledge or seeking connections with previous knowledge. On the other hand, students who have adopted a deep approach to learning are interested in the subject and study to understand and master it. These students seek out ways in which the new material connects to their daily lives and builds on their previous understanding (Kember and McNaught 2007). Fortunately, the field of cognitive science offers many effective methods to encourage deep learning. This chapter summarizes the attributes of effective learning and offers practical methods students can use to incorporate these attributes into their study strategies.

Incorporating Effective Learning Strategies

Many students have a poor intuitive feel for which study practices will improve their learning. When learning seems to be easy and rapid, they can feel a strong sense of mastery yet perform poorly on tests. However, when learning is more challenging, and therefore slower, students often feel less confident but perform better. In many of the studies cited in this chapter, the participants who had the highest confidence in their learning performed the worst. Although it may seem counterintuitive, long-term learning is greater when initial learning is more difficult (Bjork 1994a, 1994b). However, because students may not experience an immediate sense of accomplishment when initial learning is difficult, they are reluctant to replace old study strategies with more effective ones. To help students expand their repertoires of study strategies and improve their understanding of what makes a strategy effective, the following sections present "best practices" based on cognitive science research. These learning strategies are organized into two broad categories that address 1) mastering course material and 2) structuring effective study sessions.

Strategies for Mastering Course Material

In general, learning that leads to mastery of knowledge or skills is enhanced by study practices that involve less input (rereading, rewriting, reviewing) and more output (retrieval and processing of material to be learned) (deWinstanley and Bjork 2002). As shown in Table 6.1, learning activities with more output than input share the following components: generation, retrieval practice,

Table 6.1. Components of Effective Learning

Generation	Producing new information, associations, or interconnections from cues
Retrieval Practice	Retrieving information from memory
Interpretation	Describing how new information fits with what is already known
Elaboration	Viewing information in a number of different ways (variable encoding); restructuring information; considering implications of the information
Attention	Focusing on the task; not dividing attention
Metacognition	Monitoring one's own learning in a purposeful and conscious way

interpretation, elaboration, attention, and metacognition (Bransford *et al.* 1999; deWinstanley and Bjork 2002). In the following sections, these components are discussed in greater detail with specific examples of how students can integrate them into their studying.

Generation and Retrieval Practice

The act of retrieving information during testing greatly enhances learning, meaning that using one's memory actually changes it. In a study, students were divided into four groups. During each of four time intervals, they either studied or were tested on material to be learned. Students in the group that was tested during three of the four sessions performed the best on the final test (McDaniel *et al.* 2007). When asked to predict how they would do on the test, participants who only studied showed the highest confidence levels but the lowest test scores—further evidence that learners are poor at judging the effectiveness of study methods (ibid.). This "testing effect" occurs whether the studies are conducted in psychology labs (Roediger and Karpicke 2006a, 2006b) or in medical school classrooms (Larsen *et al.* 2008, 2009).

Given the robustness of the testing effect, encouraging students to use testing as a study strategy provides them with a powerful new learning tool. Instructors may facilitate "learning by testing" in several ways: 1) make an old exam available, 2) create a set of study questions, or 3) ask students to create study questions. While the last option minimizes instructor workload and gives students practice writing questions, the quality of student questions is typically low unless students have had practice with question writing (Crowe *et al.* 2008). Instructors can also use formative assessment to model learning by testing.

Interpretation and Elaboration

Information that is connected and structured has a greater likelihood of being successfully retrieved (deWinstanley and Bjork 2002). For example, mnemonics help biology students to remember cranial nerves and geology students to remember geological periods because they add structure to the information. Creating visual representations of information helps students to structure their understanding: outlines, maps, drawings, and diagrams are examples that encourage students to process information in order to organize it (ibid.). This mental processing reinforces the ability to remember the information at a later time. Creating outlines allows students to rework and reorganize class information while also providing a structure upon which to add new information. Conversely, providing students with detailed outlines in class has been shown to inhibit note taking and the learning that accompanies it (Morgan *et al.* 1988). Knowledge maps (graphical representations in which arrows depict how one factor influences another) help students understand mechanistic relationships between components of a system. The main difference between knowledge maps and concept maps is that the former emphasizes cause-and-effect relationships, while the linking phrases used in concept maps deal with a wider range of relationships between concepts (O'Donnell *et al.* 2002). Summary sheets focus on helping students see the organizational relationships between course topics and require students to use pictures, flow diagrams, annotated graphs, and tables to summarize the content and concepts discussed in each class.

When students elaborate on their current understanding rather than reread material during study sessions, they increase retention of that material (deWinstanley and Bjork 2002). In order to elaborate on a graph, students might ask themselves, "What does this graph tell me, what is the significance of the data represented, and how is this graph helping me see the relationship between two variables?" In this way, students learn to bring a similar set of questions to their interpretation of textbook figures, data tables, and animations. When making predictions, students determine whether their current understanding is accurate, as correct predictions confirm correct understanding. This is a particularly powerful technique for students to use while studying with animations: at various points during the animation, a student can stop the action and predict what will happen next (de Koning *et al.* 2010). Resuming the animation gives them their answer. Instructors can incorporate this same type of elaboration into their classes by setting up scenarios or problems related to course material and having students predict the results. This is an easy and convenient formative assessment that allows students to check their understanding and informs the instructor as to the progress of student learning. Simulations also have the potential to be a powerful learning tool for students, as they provide students with the option to change various parameters in the model and observe the consequences of those changes. However, not all simulations are equally capable of helping students learn difficult concepts. The Physics Education (PhET) group at the University of Colorado, Boulder, has done extensive research to determine the key elements of effective simulations and has developed numerous physics, biology, and chemistry simulations that are freely available at their web site, listed in Appendix B (Adams *et al.* 2008a, 2008b).

Attention

While researchers have known for decades that divided attention results in poor memory (Tyler 1969; Tyler *et al.* 1979), the general public has only recently become aware of its negative effects. Recent studies of medical professionals have shown that higher incidences of interruptions and distractions increase performance errors (Flynn *et al.* 1999; Biron *et al.* 2009; Westbrook *et al.* 2010). Although results from these types of studies are beginning to be reported by the popular media, the general public still perceives multitasking—attending to more than one task at a time—as effective and desirable. Students are particularly prone to this misconception. Given the technological age in which we live, where students are surrounded by Internet-enabled devices, persuading students to study in a distraction-reduced environment is no small challenge (Baddeley *et al.* 1984). Chun and Turk-Browne (2007) provide a useful analogy of attention that might resonate with students: they compare attention to exercise. When deciding to exercise, a person will decide if they have sufficient resources, say time and energy, to do the exercise and if so, will select a type of exercise, such as running or swimming, upon which to expend those resources. Like exercise, attention also entails resources and selection. When students decide to study, they will select a topic upon which to focus their mental resources. Multitasking while studying is problematic because it divides students' cognitive resources, reducing their ability to form precise memories (Craik *et al.* 1996).

Metacognition

Metacognition is the purposeful and conscious self-monitoring of learning. Experts in a discipline engage in metacognition as they continuously check and challenge their understanding of new concepts and paradigms. Novices, by definition, are not as adept at this process. As science novices, students can easily misjudge the depth and reliability of what they know. Wishful thinking or over-confidence can replace accurate self-evaluation when preparing for examinations. When asked at the end of an exam to predict their scores, high-performing students underestimated their performances while low-performing students overestimated (Keefer 1969). In both cases, students failed to accurately assess their level of understanding (Kornell and Bjork 2007).

Developing metacognitive skills requires that students have 1) confidence in their ability to influence their own learning, 2) a set of academic goals, and 3) numerous opportunities to practice monitoring and adapting their learning (Lovett 2008). How do students know if they are capable of learning? Interestingly, research indicates that most students hold either a fixed or fluid view of intellectual ability (Dweck and Schunk 2000). Students with a fixed view believe that everyone is born with a set intellectual potential, beyond which further improvement is not possible. They believe that this potential can be realized at any point in the educational process, from kindergarten to graduate school, and that failure is an indication that they have hit their "ceiling," which makes further work or study futile. In contrast, students with a fluid view of intellectual ability believe that the harder they work, the smarter they will become. They see intelligence as malleable, not innate, and limited only by their level of effort. For the fluid thinker, failure is merely another opportunity to learn. While a discouragingly large proportion of people (80 percent) hold a fixed view of intelligence, an individual's view of intelligence can change (ibid.). One path to change is to encourage students to set learning goals rather than performance goals. For the fixed thinker, failure to achieve performance goals (such as desired exam scores or grades) signals a lack of ability, whereas failure to meet a learning goal simply indicates lack of mastery and a need for additional study and practice.

An easy and straightforward way to facilitate student planning and goal-setting is to have students create and submit a schedule of when, what, and where they will study each week. In subsequent weeks, students indicate which study sessions they completed and which they missed. This activity allows students to assess their effectiveness in reaching stated goals. Following evaluation of their achievement, instructors can encourage students to adapt their study plans to more realistic expectations. Effective goals are directed toward mastery (understanding) rather than performance (grades) and are revisited periodically throughout the course.

A simple strategy to get students to monitor and adapt their learning is to ask them to predict their exam scores and then compare their prediction to their actual performance. A similar method has been shown to help focus students' subsequent study, resulting in improved performances on future tests (Coil *et al.* 2010). A technique called the Exam Wrapper promotes student metacognition (Lovett 2008) by asking students to complete an exam reflection sheet that directs them to describe their study strategies, analyze their mistakes on the exam, and plan their study strategies for the next test. Another simple yet effective method to help students both assess the effectiveness of their study strategies and improve their metacognition is for the instructor to include the Bloom level of each exam question on the exam key. Then students can identify the Bloom levels with which they struggle most and tailor their study strategies toward improvement at those levels. The Bloom's-based Learning Activities for Students tool (BLASt; see Appendix A) is a suggested list of learning activities that guide students to work with content at higher cognitive levels. Finally, Tanner (2012) has assembled numerous activities to enhance metacognition for both instructors and students to implement in and out of the classroom.

Strategies for Structuring Effective Study Sessions

As mentioned earlier, an interesting conundrum exists: learning that feels easy leads to an elevated perception of mastery that does not actually match reality. While the first part of this chapter presents techniques for helping students expend cognitive energy more effectively, this final section focuses on structuring more effective study sessions. Toward that goal, Bjork has identified a set of learning activities that include "desirable difficulties," which make initial learning of the material more difficult but lead to greater long-term retention. As described below, these techniques include spacing out sessions rather than "massing" them in time and space, and interleaving the study of different topics rather than studying blocks of a single topic.

Spacing across Time and Place

When preparing for exams, many students spend the day or night before the test focusing solely on the test topics: the well-known tactic of "cramming." Cognitive science calls this study pattern "massing" or "blocking." Massing or blocking study time for one topic does indeed speed up the *initial* superficial aspects of learning and thus creates a feeling of confidence. This perception of knowing the material is comforting and reinforces a sense of mastery. Students may do well if the exam is based purely on recall, but if the exam contains mostly HOC questions, performance is poor (Bjork 1979). Students generally do not retain information acquired through massing, which explains why they must restudy material for cumulative final exams, and why they have difficulty recalling information supposedly learned in prerequisite courses. Though massing allows students to feel good about their learning efforts in the moment, it undermines subsequent academic success.

Long-term recall is enhanced by spacing study efforts and testing opportunities over time (Shebilske *et al.* 1999; deWinstanley and Bjork 2002). Students who took longer intervals between study sessions achieved almost twice the recall of students who took shorter intervals (Figure 6.1; Bjork 1979). This effect was observed over a large range of time intervals and types of material, from simple recall to complex skills. The enhanced performances of students using the spaced study sessions probably result from students viewing the material in different ways and retrieving material from memory for each session, both of which enhance retention. Interestingly, when asked to predict their performance on the upcoming test, the students using spaced study sessions expressed the lowest confidence and were quite astonished to learn that they outperformed the students whose study sessions were clumped together.

Changing the location of study sessions also positively affects learning. Students who studied the same topic in different locations performed better on

Students followed either scenario 1 or 2. Students in scenario 2 had greater time between their study sessions (S) prior to completing the same test as students in scenario 1.

| Scenario 1 | | | | S1, S2, S3 S4 | Test |
| Scenario 2 | S1 | S2 | S3 | S4 | Test (Students performed best) |

Figure 6.1. Bjork's Spacing Experiment (Bjork 1979)

tests than students who repeatedly studied in the same room (Smith and Glenberg 1978). The only exception to this finding was when students could study in the room in which the exam was administered. In this case, familiarity with the test setting is thought to decrease test anxiety. (Since most students cannot study in the testing facility, this is generally not a practical option.) Like most humans, students find comfort in familiarity and thus are likely to cling to familiar study habits and places. Sharing the research findings summarized here with students may help convince them of the value of planning multiple study sessions over a greater period of time in a variety of settings.

Interleaving Rather Than Blocking Practice

Whether cramming the night before an exam or not, students tend to set aside blocks of study time to focus on a single topic. In a study where students learned to recognize the work of different painters either by viewing all paintings of one artist at a time (blocking) or by alternating paintings from all the artists (interleaving), recall by students in the interleaving group was far superior (Figure 6.2; Kornell *et al.* 2009). Taylor and Rohrer (2010) found similar results when students were learning to solve four different kinds of math problems. These results suggest students can maximize learning by alternating rather than blocking study topics.

Study strategies that incorporate "desirable difficulties" by requiring students to process information and move out of their comfort zone maximize learning. However, it is important to realize that not all learning difficulties are considered desirable. For instance, studying complex material while fatigued leads to diminished, not increased, learning. Similarly, the field of cognitive load theory emphasizes that working memory is limited in its capacity and that instructional strategies that overload this capacity lead to less effective learning. Identifying which difficulties are "desirable" is an area of ongoing research in the learning sciences.

Each number represents a painting by a given artist. For example, 1 would be all paintings by Monet.

Blocked: 1 1 1 1   2 2 2 2   3 3 3 3   4 4 4 4
Interleaved: 1 2 3 4   2 1 4 3   4 3 2 1   3 4 1 2   (Best test performance)

Figure 6.2. Blocked versus Interleaved Practice (Kornell *et al.* 2009)

Conclusion

The goal of most instructors is to encourage students to move away from a focus on grades and toward a mastery and appreciation of the discipline. The challenge is how to help students shift from a superficial to a deep approach to learning. The "illusion of knowing" is very alluring to students, as it provides a psychological sense of ease at a time when they feel threatened by their lack of understanding of complex subject matter. Fortunately, the cognitive sciences as well as discipline-based education research (see Chapter 7) have identified key elements of successful learning. As instructors, we can model the components of effective learning in the classroom and provide opportunities for students to practice generation of ideas, retrieval of information, interpretation of course material, elaboration of concepts, focused and attentive behaviors, and reflective practices. We can also explicitly share learning research with our students and remind them that when learning is too easy or comfortable, it is usually not very effective. Better understanding of the power of effective study strategies offers students viable alternatives to flash cards and highlighting the text with multiple markers.

Students also need to realize that as they progress through their academic careers, greater emphasis will be placed on working at the higher cognitive levels of analyzing, synthesizing, and evaluating information. They will be expected to develop robust conceptual frameworks and mental models that are foundational to the discipline. Just as it is important for instructors to align assessments with intended learning outcomes and with teaching methods, it is also important that students adapt and align their study strategies with the level at which they are expected to perform. Change is never easy for either students or instructors, but deliberate practice of the key elements of learning can ultimately lead to success in the classroom.

Appendix A: Bloom's-based Learning Activities for Students (BLASt)

Adapted from Crowe *et al.* 2008

Bloom Level	Individual Activities*	Group Activities*
Knowledge (LOCS)	• Practice labeling diagrams • List characteristics • Identify biological objects or components from flash cards • Quiz yourself with flash cards • Take a self-made quiz on vocabulary • Draw, classify, select, or match items • Write out the textbook definitions	• Check a drawing that another student labeled • Create lists of concepts and processes that your peers can match • Place flash cards in a bag and take turns selecting one for which you define a term • Do the above activities and have peers check your answers
Comprehension (LOCS)	• Describe a biological process in your own words without copying from another source • Provide examples of a process • Write a sentence using the word • Give examples of a process	• Discuss content with peers • Take turns quizzing each other on definitions and your peers will check your answer
Application (LOCS/HOCS)	• Review each process you have learned and then ask yourself what would happen if you increase or decrease the activity of a component in the system • If possible, graph a biological process and create scenarios that change the shape or slope of the graph	• Practice writing out answers to old exam questions on the board and have your peers check that you don't have too much or too little information in your answer • Take turns teaching your peers a biological process while the group critiques the content
Analysis (HOCS)	• Analyze and interpret data in primary literature or a textbook without reading the author's interpretation and then compare the author's interpretation to your own • Analyze a situation and then identify the assumptions and principles of the argument • Compare and contrast two ideas or concepts • Create a map of the main concepts by defining the relationships of the concepts using one- or two-way arrows	• Work together to analyze and interpret data in primary literature or a textbook without reading the author's interpretation and defend your analysis to your peers • Work together to identify all of the concepts in a paper from the primary literature or a textbook chapter, create individual maps linking the concepts together with arrows and words that relate the concepts, and then grade each other's concept maps
Synthesis (HOCS)	• Generate a hypothesis or design an experiment based on information you are studying • Create a model based on a given data set • Create summary sheets that show how facts and concepts relate to each other • Create questions at each Bloom level as a practice test and then take the test	• Propose a hypothesis about a biological process and design an experiment to test the hypothesis, then have peers critique the hypotheses and experiments
Evaluation (HOCS)	• Provide a written assessment of the strengths and weaknesses of your peers' work or understanding of a given concept based on previously determined criteria	• Provide a verbal assessment of the strengths and weaknesses of your peers' work or understanding of a given concept based on previously described criteria, and then have peers critique your assessment

*Students can use the individual activities, group activities, or both to practice their ability to think at each level of Bloom's Taxonomy.

Appendix B: Animations

Physics Education (PhET) group at University of Colorado, Boulder: http://phet.colorado.edu/en/simulations/category/new

III

ASSESSMENT BEYOND THE CLASSROOM

The final section of this book explores the uses of assessment beyond the classroom. The first chapter serves as a primer on evaluating the impact of teaching as well as on conducting research that advances our understanding of best practices in science education. The final chapter provides a guide with the necessary resources for running a workshop on assessment to pass on what you have learned to your colleagues.

7

Assessing the Effectiveness of Teaching Innovations

> Anecdotal thinking comes naturally; science requires training.
>
> —Michael Shermer

As researchers and educators, science faculty should be able to assess the effectiveness of teaching innovations designed to enhance student learning. Using their scientific training, faculty can ask questions, conduct experiments, and collect and analyze data in order to draw conclusions. Instead of basing perceptions of effective teaching on traditional norms and personal experience, faculty can engage in action research (AR) by applying their scientific skills to answering questions about student learning or teaching efficacy. When evidence generated from AR moves into the realm of education research and publication, it becomes known as "discipline-based education research," or DBER. This chapter offers a general overview of methods and tools for AR and DBER.[1] Another book in the Scientific Teaching series, *Discipline-Based Education Research: A Scientist's Guide,* presents a more in-depth look at many of the topics highlighted here

---

[1] To avoid confusion we have not used the somewhat ambiguous term Scholarship of Teaching and Learning (SoTL), which is sometimes used to mean AR but can also include DBER.

(Slater *et al.* 2010). Other resources also offer more detailed treatments of these topics (for example, Wood 2009 and NRC 2012b).

Scientific Teaching Using Action Research

The same scientific approaches that are applied in the laboratory and in the field can also be applied to the classroom. In the practice of "scientific teaching" (Handelsman *et al.* 2007), faculty bring the art of research into their classrooms by reflecting on and improving their teaching after collecting and analyzing evidence about student learning. Rather than making assumptions about their students' learning, they use AR to collect evidence to support or reveal inadequacies in their pedagogical practices and thereby strive to improve student learning outcomes. Scientific teaching assumes that faculty are methodical in their approach, employing best practices established by cognitive research on teaching and learning. Not only do faculty apply science to teaching, but they bring the discovery process of science into student learning with the hope that students will be excited by both the content and process of science.

Faculty can use the tools outlined in this book to engage in scientific teaching by applying scientific approaches to measuring success of instruction and student learning in meaningful ways. Section I of this book emphasizes the importance of using backward design and Bloom's Taxonomy to provide a framework for assessing instructional materials and student learning. Section II first highlights how assessment drives learning and then offers suggestions for improving classroom assessments and student preparation for exams. Section III outlines practices and strategies for instructors interested in applying educational research to their teaching. Although we emphasize AR throughout the text, the methods and resources are also applicable to DBER.

Action Research in Practice

AR provides evidence about teaching effectiveness that can then be used for multiple purposes—as documentation for tenure or review, to inform pedagogical decisions, or to determine the extent to which students are achieving intended outcomes, for example. But the principal purpose of AR is to apply the lessons learned to subsequent actions in one's own classroom. AR allows the instructor to evaluate the success of an instructional intervention, with the intent to take action on the results—often immediately. Usually AR is done by the practitioner, rather than an outside observer or evaluator.

Qualitative versus Quantitative Data

Implementing AR requires a strategic plan for collecting the kind of data that will best help answer a question. Given that most natural scientists collect and use quantitative data, they are often comfortable selecting this approach for studies in the classroom. Exam scores, students' incoming GPAs, learning gains measured by pre- and post-testing, and student surveys based on a Likert scale (a rating scale in which responses are scored along a range) are examples of quantitative data that faculty often gather. But while quantitative data are useful in determining *what* a student knows, they are generally not helpful for understanding *how* or *why* a student knows it. To learn how students approach balancing chemical equations or why they think the majority of a plant's biomass comes from something other than air, the instructor needs to gather some form of qualitative data that allows students to describe their reasoning. Interviews, written papers, observations, or audio-visual recordings of student work can tell an instructor much more about the how and why of student thinking than most quantitative classroom assessments.

Experienced AR practitioners often collect both quantitative and qualitative data, and in some cases even quantify qualitative data through coding of responses. For example, designers of quantitative instruments, such as the GRE or MCAT, use both student scores and interviews to improve test items. Students first take the test, which generates quantitative data about each question, and then they discuss how and why they selected their answer for each question, providing qualitative data about their reasoning. The information gathered by the observer helps the test designer refine the test questions so that students don't get the right answer for the wrong reason or miss a question because it is unclear. Faculty can also use this approach to improve their own exams. This is just one way in which a mixed-methods approach can provide a more complete understanding of a complex process—in this case, student reasoning.

It is also not uncommon for action researchers to collect qualitative data for the purpose of better understanding students' attitudes or beliefs about a particular topic without a specific hypothesis in mind. For example, a geology instructor may want to learn *before* teaching the topic what students think about how the Earth formed. In this case the instructor would collect qualitative data, sort through the responses to find common student misconceptions, and subsequently change teaching plans to address those misconceptions. If instructors do this several times for subsequent versions of the same course, they may identify clear modes of thinking, or themes, among introductory geology students. This kind of approach is referred to as "grounded theory," and may generate useful information that can be used to design future hypothesis-driven research (Corbin and Strauss 1990).

Collecting qualitative data for generating quantitative data, or mining qualitative data for discovery, is referred to as "interpretive research" (Erickson 1986). This chapter is meant to provide a starting point for AR by building on previous resources, so although it touches on some aspects of gathering and assessing qualitative data, the details of interpretive research are beyond the scope of this chapter. For more on interpretive research designs, see Chapter 5 in Slater *et al.* (2010).

Selecting the Appropriate Instrument for Collecting Data

Once instructors determine what kind of data they want to collect, they need to identify or create an instrument or method for collecting it. Most AR studies aim to gather data about changes in content knowledge, conceptual understanding, attitudes, beliefs, self-perceptions, or science process and reasoning skills, to name a few. Qualitative or quantitative data can be collected using a variety of methods and instruments such as interviews, concept inventories (defined in the next paragraph), or attitude surveys (see Appendix A for more on concept inventories and attitude surveys). In some cases, it may be necessary to use more than one assessment instrument or method. For example, an instructor implements group problem-solving activities and wants to determine if students' conceptual understanding and perceptions of science as a process (rather than a collection of facts) improve. To collect these kinds of data, the instructor might use a concept inventory as a pre- and post-test to measure changes in conceptual understanding, coupled with interviews or an attitude survey to assess changes in students' attitudes toward science. After the instructor decides how to collect the data, DBER literature can be used to find the appropriate instrument for the job.

AR scholars often rely on the use of existing assessment instruments that have been shown to be valid and reliable. Validated instruments measure what they are supposed to measure, and reliable instruments consistently distinguish between individuals with disparate abilities (Alias 2005). Many valid and reliable instruments are published in the DBER literature and often are in a multiple-choice format, making them easy to administer and grade for large numbers of students. These various types of validated instruments have been designed for assessing student content learning, skill acquisition, or attitude change in science disciplines. The type of instrument called a concept inventory is a validated multiple-choice assessment tool that has been used in AR by faculty in many science disciplines to document normalized learning gains within a class (Marx and Cummings 2007). Concept inventories have also been used to document differences in teaching within an institution (Marbach-Ad *et al.* 2010; Smith *et al.* 2008; Marbach-Ad *et al.* 2009). Validated instruments are available for measuring science process and reasoning skills as well as attitudes about science and research. Appendix A lists some of the validated instruments currently available for different science disciplines.

In addition to using multiple-choice instruments such as concept inventories, investigators may want to use assessment tools based on free-response questions. Because of the subjective nature of scoring free-response questions, these types of tools tend to give less reliable, albeit often richer, results than multiple-choice instruments. To more reliably evaluate answers to free-response questions, one can design a validated rubric with high inter- and intra-rater reliability (see Box 7.1 for definitions). Chapter 3 discussed how rubrics can be used to assess student work and enhance student learning, but did not highlight how to use rubrics for AR. Faculty engaged in AR will need to achieve intra-rater reliability from a validated rubric so they know that the results achieved from a particular teaching innovation are based on the innovation, not on inconsistent use of the grading rubric. Guidance for designing validated rubrics is provided in Appendix B.

Selecting a Research Format

In contrast to most basic biological research, AR studies are often only quasi-experimental because they lack randomization of subjects and often lack control groups consisting of students of academic abilities equivalent to those in the test group. Fortunately there are research design strategies that can help overcome this limitation.

Research can be designed to compare differences in knowledge, attitude, or skills either between the same group of students at the beginning and end of a course (pre- and post-test format) or between different groups of students in subsequent years of a course taught by the same instructor (sequential course format). Alternatively, internal comparisons within a class (cross-over format) can be used to measure changes of specific groups, so that instructors can evaluate their teaching methods and adopt strategies that help all students improve. The following descriptions outline three common research formats for investigating teaching and learning. All three formats involve comparison of pre- and post-test results, so it should be noted that such comparisons only document change in students' performance or attitudes; the cause of that change may not necessarily be due to an investigator's teaching interventions, but rather due to other outside factors, such as a concurrent research experience. Therefore a mixed-methods approach, using student interviews or surveys in addition to pre- and post-tests, should be employed to determine what might be causing observed change.

Pre- and post-test format is used to assess learning gains of individual students or all students in the class. Students are tested at the beginning of a course and again at the end, with differences in performance calculated as learning gains. This format controls for differences in the incoming preparation levels or academic abilities of students. Measuring learning gains may show that changes in teaching practices or other course interventions improved students' learning of content or skills. One consideration in using this design is the test itself. The same test, or a different test with alternate question forms (also referred to as isomorphic questions) that assess the same concepts, should be given at the beginning and end of the course or instructional unit. The test shouldn't use terminology that incoming students might be unfamiliar with, should have questions that can discriminate between students' abilities, and should be sufficiently difficult that no student can answer all the questions (see section on Calculating Learning Gains from Pre- and Post-Tests).

Sequential course format uses pre- and post-testing to compare different groups of students in the same course over subsequent semesters. This allows an instructor to determine whether a particular teaching intervention can consistently elicit the same level of student achievement with different groups of students. As mentioned above, the pre- and post-tests must be sufficiently difficult to discriminate between students' abilities. The sequential course format may also help instructors determine if new teaching interventions can help to iteratively improve students' learning of a particular topic or skill. For example, a geology instructor teaches about minerals in her course and over several years has improved student learning through a variety of in-class activities. She wants to test whether the addition of yet another graded in-class activity would further improve students' abilities to learn minerals. She adds the new activity in two different semesters of the same course and compares the learning gains of students in the new version of the course to the learning gains of students who took the course in previous semesters. She finds that the new activity does not enhance their learning of minerals, so she works to create a different activity to test the next time she teaches the course.

Cross-over research format is another approach that helps to control for differences in students' abilities within a class. Students are divided into groups A and B, with group assignments randomized if possible. For a particular teaching intervention, group A serves as the treatment group while group B serves as the control group. The groups are then switched for a different topic that uses the same pedagogical approach. Student learning is assessed with pre- and post-tests, perception surveys, interviews, or another method to determine if the new approach was effective for improving learning. With this format, it is important to match the cognitive challenge of the two teaching modules so that one can draw valid conclusions about the efficacy of the teaching intervention.

Calculating Learning Gains from Pre- and Post-Tests

Learning gains for individuals or an entire class can be calculated from performances on pre- and post-tests. The pre- and post-tests can be identical or isomorphic, and faculty who use either should be aware of the caveats associated with both methods; the Research Methods Knowledge Base is an excellent resource (Trochim 2013). Data from both identical or isomorphic pre- and post-tests can be used to calculate learning gains using the following formulas (Hake 2001).

The normalized gain for a single student, <g>, is calculated as the percent gain divided by the maximum possible gain:

$$<g> = (\%\text{post-test} - \%\text{pre-test}) / (100 - \%\text{pre-test})$$

Normalized gains for all the students in the course, $<g_{class}>$, is calculated as the average gain divided by the maximum possible actual average gain using the same formula, except that %(post-test) and %(pre-test) are the final and initial class averages, respectively.

When analyzing results from pre- and post-tests, it may be necessary before calculating normalized gains to clean up the data as follows: 1) remove students who earned perfect scores on the pre-test because an improvement of that student's performance is beyond the scope of measurement for the assessment instrument; 2) remove students whose post-test scores are less than their pre-test scores, as $<g>$ has no meaning under that circumstance; and 3) remove students who score 100% on the post-test because regardless of their pre-test score, they will have a $<g>$ of 1, which is not informative. However, even in these special cases it is possible to evaluate student performance as normalized change instead of gains (Marx and Cummings 2007). If an assessment is sufficiently difficult, most of these problems can be avoided. Although there may be a few caveats with this approach, learning gains can often inform instructors about student learning and their teaching practices throughout the course or program.

Moving from Action Research into Discipline-Based Education Research

Faculty across science and non-science disciplines employ AR to collect accurate information with which to increase the effectiveness of their teaching and further develop their pedagogical expertise. When faculty go beyond AR studies and conduct education research with the intent to publish their findings, they are engaging in DBER. This kind of research draws on findings from AR practitioners, educational psychologists, and cognitive researchers to "investigate learning and teaching that is informed by an expert understanding of disciplinary knowledge and practice" (NRC 2012b). The following sections of this chapter build upon the methods and techniques of AR by outlining some key DBER practices and strategies for instructors wanting to make the transition from AR to DBER, and provide important references for getting started. Given that DBER is often hypothesis-driven, we introduce DBER practices through the scientific approach: identifying a good research question, designing your study, selecting or creating the correct tools for collecting meaningful data, and considering various approaches to data analysis. At the end of the chapter we discuss the ethical issues and appropriate steps one must take prior to conducting research intended for publication.

Identifying a Good Question and Designing Your Study

The first step in DBER is to pose a question that addresses issues of concern for instructors and education scholars in a specific science discipline. The findings should inform a broad scientific community rather than an instructor for a single class, department, or institution. DBER research questions can typically be addressed only by instructors within that specific discipline, since teachers in a particular field have pedagogical content knowledge (Bransford *et al.* 1999) as well as insight into the challenges their students encounter when learning discipline-specific content or skills. Instructors can mine their classroom experiences for pertinent questions about teaching and learning in their discipline, read the DBER literature, and link their investigations to what is known in the context of an established theoretical framework. Identifying a theoretical framework—the assumptions and approaches the investigator takes into the research—is important to identify if the work is fit to be published. Reading the DBER literature helps in learning more about the kinds of research questions that are investigated and how they are tied to theoretical frameworks. Numerous science education journals serve as excellent resources for designing DBER studies, and the University of Central Florida Libraries has compiled an outstanding list of relevant journals (http://www.fctl.ucf.edu/ResearchAndScholarship/SoTL/journals).

Selecting or Creating the Correct Tool for Collecting Meaningful Data

As mentioned previously in this chapter, it is often possible to find an appropriate validated instrument prior to beginning an AR or DBER project. However, if a suitable tool does not exist, one can be created, although this is not a trivial undertaking. Developing an instrument requires establishing its validity, followed by iterative piloting to show that it is reliable. Most importantly, the test needs to be designed in such a way that the resulting data can elucidate patterns that help the investigator test his or her hypothesis. There are many kinds of tests, each with a different purpose. For example, one test might be designed to measure very specific content knowledge of upper-division students using multiple true/false questions (see Chapter 5), whereas another might be designed to measure change in students' process skills throughout their undergraduate education using standard multiple-choice questions. From the outset, the overall approach to designing each of these kinds of tests and the way in which the data is analyzed would be different. Reviewing the literature on the development of tests can help to highlight these differences. Appendix A provides a list of validated instruments, some of which, such as the Force Concept Inventory (FCI), have been used extensively for more than a decade (Hestenes *et al.* 1992). Since the FCI was created, many validated instruments (listed in Appendix A) have been designed for use in Science, Technology, Engineering, and Mathematics (STEM) and serve as excellent resources for instructors wanting to develop their own.

Although no two validated instruments are the same, they are all designed to achieve the qualities of a well-made assessment tool. These qualities, such as item discrimination or content validity, are often discussed in the literature and can be confusing to someone who is not familiar with their use. Earlier in this chapter we defined validity and reliability in the general sense, but it is useful to know that there are different kinds of validity and reliability. Although it is beyond the scope of this chapter to show the many ways in which qualities are determined or how they are calculated, we have provided a list of definitions (Box 7.1) that may be helpful to anyone reading the instrument development literature for the first time.

Approaches to Data Analysis

There are many ways to analyze results from DBER studies. Since many of the techniques overlap with what faculty already do in their own scientific research, we will only briefly highlight two important methods that may be overlooked: calculations of effect size, and item response theory.

Effect Size

Effect size is a measure of differences between groups that emphasizes the magnitude, rather than the statistical significance, of the difference (Nakagawa and Cuthill 2007). Effect size helps determine whether the difference seen between treatment and non-treatment groups is substantial enough to merit a change in practice. If we calculate only the statistical significance between the treatment and non-treatment groups, and the sample size is large enough, we would likely find a significant difference between groups. Does this mean we should change our teaching based solely on this calculation? If the effect size is sufficiently large

Box 7.1 Definitions Associated with Validated Instruments

(Carmines and Zeller 1979; Alias 2005; Cook and Beckman 2006; Streveler *et al.* 2011)

Item: An individual question or task to which a student will respond.

Item Difficulty: The percentage of students answering the item correctly (a confusing but unfortunately established convention; note that the easiest items have the highest item difficulty).

Item Discrimination: The item's ability to discriminate between students of different academic ability—those who have mastered the material and those who haven't. The students with mastery of the content are more likely to answer an item correctly.

Reliability: Addresses the consistency of a set of measures and is based on whether the test results are repeatable and not subject to random error. Values for reliability are typically reported as a correlation coefficient resulting from a statistical test: Cronbach's alpha (Cronbach 1951) for tests with non-dichotomous (multiple possible answers) items or Kuder-Richardson Formula 20 (K-R20) for dichotomous items (Kuder and Richardson 1937). The higher the correlation coefficient, the greater the reliability.

> **Alternative Form Reliability:** A student will have relatively the same result when they take alternative (isomorphic) forms of the same test at two different times because the different items assess the same content or skills at the same cognitive level.
>
> **Internal Consistency:** Measures whether pair-wise items on a test that should measure the same content, or latent variable, actually do so.
>
> **Inter-Rater Reliability:** The consistency of scoring free-response items by different graders.
>
> **Intra-Rater Reliability:** The consistency of scoring free-response items by the same grader at different times.
>
> **Test–Re-test Reliability:** A student or group of students will score similarly if they take the same test at two different times without intervention.

Validity: Addresses whether the test measures what it is supposed to measure.

> **Construct Validity:** The *item* measures what it is intended to measure (requires affirmation by experts).
>
> **Content Validity:** The *test* presents the key concepts and misconceptions for a given domain as affirmed by experts. Content validity increases as the number of relevant items on the test increases.

Face Validity: The test has the appearance of measuring what it intends to measure, thereby motivating students to do their best. The assumption is that students performed to the best of their abilities on the test.

and there is statistical significance, then the treatment, or teaching innovation, has a real impact on student learning. The scenario below describes a situation in which effect size, not just significance, is used in data analysis.

> An instructor hypothesizes that students learn experimental design best by designing and then doing experiments in a laboratory setting. She teaches a large introductory biology class of over 700 students who are taking lab sections in groups of 25. She assigns students in half the lab sections to design and do their experiment each week in the lab, while the other half designs the same experiments on paper and doesn't do the experiment. After four classes, all her students take a test to measure their ability to design an experiment. She finds that the means for the two groups are significantly different but there was a lot of variation within groups, so she calculates effect size. She realizes the effect size is small and is no longer confident in the data— she still wonders whether doing lab experiments helped students learn experimental design. She decides to continue the research the following semester but will give the students a pre-test to determine who already knows how to design experiments before labs so she can focus her study on those students who have the most to gain from the practice.

Calculating effect size, the difference in two means divided by a measure of the standard deviations, can help the instructor determine if a change in instruction is warranted. Effect size is usually calculated as Cohen's d, where *n* and *s* are numbers of students in groups 1 and 2, respectively (http://www.philender .com/courses/intro/notes/effect.html).

$$d = \frac{\text{mean of experimental group} - \text{mean of control group}}{\text{pooled standard deviation } (p_s)}$$

$$p_s = \frac{\text{sqrt}(((n_1 - 1)s_1^2 + (n_2 - 1)s_2^2)}{(n_1 + n_2 - 2))}$$

Although there is not an exact cutoff for what is considered a meaningful effect size, Cohen described 0.2 as "small," 0.5 as "medium," and 0.8 as "large" on a relative scale (Cohen 1969). Thus, one might consider changing a practice based on a medium or large effect size.

Item Response Theory

For more than fifty years there have been two statistical frameworks that test designers have used: classical test theory and item response theory (IRT). Under each of these frameworks are many different models that the measurement community uses to develop tests with specific goals in mind. Both frameworks are currently used, but many psychometricians have shifted from using classical test theory to using IRT (Hambleton and Jones 1993; Baker 2001). Classical test theory relies on test-score models, where the test scores for a given test group are analyzed and items are modified on the basis of the group results. As such, there is great emphasis on finding and using the most suitable group for test development in classical test theory. In contrast, IRT focuses on the analysis of each item in relation to an examinee's ability, and is considered to be test-independent (Hambleton and Jones 1993). Examinees come to the test with ability levels observable only through statistical analysis of their performance on each test item (Lord 1953). For each item, someone of a given ability has a certain probability of answering the question correctly (Hambleton and Jones 1993). The ability level is very useful since a test can be designed after an initial pilot by first identifying the overall profile the test developer wants to have—remembering that a test should maximize discrimination between the ability levels of a particular group of examinees—and then building the test with items that will give it that particular profile (ibid.). IRT commonly incorporates a one-parameter logistic model known as the "Rasch model" (Rasch 1960; Hambleton and Jones 1993). While IRT is fairly complex, there are now several software packages that support it and several excellent references that explain and compare it with classical test theory (Hambleton and Jones 1993 is a good starting point).

Considering Ethical Issues Related to Human Subjects

DBER must conform to the federal guidelines for research with human subjects, regardless of whether the subjects are undergraduates, graduate students, post-doctoral fellows, or faculty. Prior to the start of any research, an Internal Review Board (IRB) committee must review and approve the research design. Therefore, early in the design process it is valuable to get feedback from a representative of the IRB committee at the institution where the research will be conducted. While many DBER studies are exempt from a variety of human subject protocols, only IRB officials can grant exemption status for research through an official review process. All federally funded science education research requires IRB pre-approval, and peer-reviewed journals will not accept manuscripts of studies that have been done without it.

Conclusion

The information presented in this chapter provides a basic foundation for getting started in AR and DBER, highlighting some important topics for engaging in this kind of work. Although science education research may sometimes be done with different methods than those used for scientific studies, the intent is the same: to discover something new. Whether instructors use diagnostic tools, interviews, or a mixed-methods approach, science education research requires creativity, problem-solving, and many other skills common to science. Therefore, the transition from teaching science to doing scholarly research in science education can be relatively easy with some guidance. As in any type of research, reviewing the literature and formulating a good research question can help to design a meaningful study that may advance knowledge about teaching and learning for the larger community. Science education research requires that the study design be well thought-out and use best practices established by cognitive researchers, psychometricians, and DBER practitioners. Our brief introduction to these methods is just a starting point for investigations on the teaching and learning of science.

Appendix A: Validated Instruments for Discipline-Based Education Research in Higher Education Science, Technology, Engineering, and Math (STEM) Fields

Concept Inventories in Astronomy

Astronomy Diagnostic Test (ADT)	Hufnagel 2002
Lunar Phases	Lindell and Olsen 2002
Light and Spectroscopy Concept Inventory	Bardar *et al.* 2007

Concept Inventories in Biology

Genetics Concept Assessment (GCA)	Smith *et al.* 2008
Genetics Literacy Assessment Instrument 2 (GLAI-2)	Bowling *et al.* 2008
Conceptual Inventory of Natural Selection (CINS)	Anderson *et al.* 2002
Biology Literacy (http://bioliteracy.net/)	Klymkowsky *et al.* 2010
Diagnostic Question Clusters: Biology	Wilson *et al.* 2006
	D'Avanzo 2008
Host-Pathogen Interactions (HPI)	Marbach-Ad *et al.* 2009
Introductory Molecular and Cell Biology Assessment (IMCA)	Shi *et al.* 2010

Concept Inventories in Chemistry

Chemistry Concept Inventory	Mulford and Robinson 2002
	Krause *et al.* 2003

Concept Inventories in Engineering

Engineering Thermodynamics Concept Inventory	Midkiff *et al.* 2001
Heat Transfer	Jacobi *et al.* 2003
Materials Concept Inventory	Krause *et al.* 2003
Signals and Systems Concept Inventory	Wage *et al.* 2005
Static Concept Inventory	Steif *et al.* 2005
Thermal and Transport Science Concept Inventory (TTCI)	Streveler *et al.* 2011

Concept Inventories in Geoscience

Geoscience Concept Inventory (GCI)	Libarkin and Anderson 2005

Concept Inventories in Math and Statistics

Statistics Concept Inventory (SCI)	Allen 2006
Calculus Concept Inventory (CCI)	Epstein 2005

Concept Inventories in Physics

Force Concept Inventory (FCI)	Hestenes *et al.* 1992
The Force and Motion Conceptual Evaluation (FMCE)	Thornton and Sokoloff 1998
Thermal Concept Evaluation	Yeo and Zadnick 2001
Brief Electricity and Magnetism Assessment (BEMA)	Ding *et al.* 2006
Conceptual Survey in Electricity and Magnetism (CSEM)	Maloney *et al.* 2001

Measuring Students' Science Process and Reasoning Skills

Rubric for Science Writing	Timmerman *et al.* 2010
Student-Achievement and Process-Skills Instrument	Bunce *et al.* 2010

Measuring Students' Attitudes About Science, Research, or Study Methods

Colorado Learning Attitudes about Science Survey (CLASS)	http://www.colorado.edu/sei/class/
Revised Two-Factor Study Process Questionnaire	Biggs *et al.* 2001
Student Assessment of Their Learning Gains (SALG) Instrument	http://www.salgsite.org/
Survey of Undergraduate Research Experiences	Lopatto 2004
Views About Sciences Survey (VASS)	Halloun and Hestenes 1998

For a more complete list of validated instruments, visit the Field-tested Learning Assessment Guide website (http://www.flaguide.org) or see the Mental Measures Yearbook or other resources at the Buros Institute of Mental Measurements (http://buros.org/mental-measurements-yearbook).

Appendix B: Creating a Valid and Reliable Rubric

If an assessment tool is based on free-response questions, creating a valid and reliable grading rubric will make scoring students' responses more consistent and reliable whether for the same grader or among different graders—referred to as intra- or inter-rater reliability, respectively. To create a valid rubric, experts in the field must agree on what the rubric assesses and what qualifies as an exemplary answer. A descriptive and valid rubric reduces possible bias in evaluating student performance and focuses the rater on the key behaviors and levels of performance instead of unrelated student characteristics, such as demographics.

Once a rubric is created, to maximize inter-rater reliability it is necessary to show that all raters interpret and implement the rubric in a similar manner and interpret students' answers the same way. To establish inter-rater reliability, expert raters first score a small number of responses (approximately five), discuss their results, and modify the rubric if necessary. Once the raters' scores are relatively similar (small standard deviation), each rater independently scores another small set of responses (five to ten) without discussion. Based on this last set of scores the inter-rater reliability is calculated using an appropriate statistical software program such as ReCal (Freelon 2010; http://dfreelon.org/utils/recalfront). Note that when using these statistical tools, it is important to identify the number of raters and type of data (Alias 2005; Freelon 2010), as these parameters will determine the appropriate type of statistical test. If Cohen's Kappa is used as the measure of inter-rater reliability, values between 0.40 and 0.59 are considered moderate agreement, while values of 0.60 to 0.80 show substantial agreement; any score above 0.80 is considered outstanding (Landis and Koch 1977). If the group does not achieve acceptable inter-rater reliability, the rubric is discussed again and the raters score yet another small set of responses. This process is done iteratively until the group achieves a reasonable inter-rater reliability value. In a similar fashion, a faculty member who scores the same group of student responses at different times can use those scores to calculate his or her intra-rater reliability.

8

Assessment Workshop

Disseminating What You Have Learned

> You cannot hope to build a better world without improving the individuals. To that end each of us must work for [our] own improvement and at the same time share a general responsibility for all humanity, our particular duty being to aid those to whom we think we can be most useful.
>
> —Marie Curie

Educational reform typically begins with, but cannot be sustained by, a single person. The goal of this chapter is to provide a framework, guidelines, and resources for disseminating the key elements of using assessment in the college science classroom. The workshop presented here is meant to help individuals who wish to pass on what they have learned from this book to their colleagues.

Assessment Workshop

This chapter is based on the assessment workshop presented by the authors at the National Academies Summer Institute on Undergraduate Biology Education. The workshop models backward design and the activities engage participants in the assessment strategies that are presented in this book.

The workshop should be suitable for a variety of target audiences, including faculty from any type of post-secondary institution as well as future faculty ranging from undergraduates to post-doctoral fellows. The units can be presented together as a single workshop (approximately two hours) or as a series of smaller workshops that address the individual topics of backward design, summative assessment, and formative assessment (see the overview in Table 8.1 for estimated times for individual activities). A single event may be a more time-efficient dissemination method for the presenter, while a series of workshops may be more beneficial for participants, as strategies for improving their teaching will be revisited several times. The smaller units could also be incorporated into undergraduate or graduate-level classes, departmental faculty meetings, seminar series, or journal clubs.

I. Preparation

Complete the following steps in preparation for the workshop. Refer to the indicated chapters for key information for both the background and the mini-lecture sections of the activities.

  A. Read or review Chapters 1–5 of this book.

  B. Review the workshop activities described below.

  C. Make copies of the resources needed for each participant.

  D. Ask participants to bring a copy of an exam from their course (or a colleague's). Alternative: bring copies of an exam from your own course (or a colleague's).

Materials Needed

For large-group activities:

▶ Flip chart, whiteboard, chalkboard, or computer and projector
▶ Computer with video capabilities and projector (for showing *A Private Universe*)
▶ Markers
▶ Nametags or table tents

For small-group or individual activities:

▶ Sheets of paper or index cards

Before the workshop:

▶ Copies of this book or access to Chapters 1–5 (optional but suggested)
▶ Exams from courses (can be brought by participants or facilitator)
▶ *A Private Universe* video (http://www.learner.org/resources/series28.html)

II. Overview and Goals

The goals of this workshop are to help participants 1) implement backward design and 2) construct meaningful learning experiences for their students using formative and summative assessment. Table 8.1 provides a list of workshop activities designed to help participants achieve the intended learning outcomes for the workshop. Sharing all or part of this overview at the beginning of the workshop will make the intended learning outcomes explicit to participants, thereby modeling good scientific teaching practices and enhancing the learning experience.

Table 8.1. Overview of Assessment Workshop

TOPIC	SUGGESTED ACTIVITIES	INTENDED LEARNING OUTCOMES Participants will…
Welcome and Overview	Introduce the intended learning outcomes and schedule of the workshop (2 min to introduce learning outcomes and ~15 sec/person for participant introductions).	know the schedule and expectations for the workshop.
Backward Design	1. Pre-assessment brainstorm (3 min)	share attitudes about and prior experiences with assessment.
	2. Think-pair-share: explicit intended learning outcomes (15 min)	be able to create instructional materials that use backward design and are oriented toward the intended learning outcomes.
	3. Mini-lecture: backward design (5 min)	be able to align the intended learning outcomes, formative assessments, and summative assessments using Bloom's Taxonomy.
Summative Assessment	1. Mini-lecture: summative assessment and alignment (5 min)	compare and contrast how formative and summative assessments drive student learning.
	2. Think-pair-share: "Blooming" summative assessment questions (25 min)	categorize summative questions according to Bloom's Taxonomy.
Formative Assessment	1. Mini-lecture: formative assessment (5 min)	describe the relationship between formative assessment and active learning.
	2. Think-pair-share: gauging learning (15 min)	build a toolbox of methods to engage students' previous knowledge, confront misconceptions, and help students construct new knowledge using formative assessments.
	3. Formative assessment in action (30 min; depends on number of activities used)	
	4. Reflection: feedback on formative assessment (5 min)	
Summary	Restate key points of the workshop (2 min)	summarize the key points of the workshop.
Exit Assessment	Participants write a one-minute paper on each of the following prompts (2 min): *What is the most important function of assessment? Did your answer change? If so, how and why?* *How can different forms of assessment drive student learning?*	evaluate the functions of assessment and explain how different forms of assessments drive learning.

III. Workshop Activities

A Note on Facilitation

Below are instructions for leading the workshop activities. Since there is always more than one way to teach a topic effectively, presenters should feel free to substitute other activities that will help participants reach the intended learning outcomes. Whether designing other activities or using those provided, it is important to remember that participants must be actively engaged in constructing their own understanding and finding their own solutions. If participants are largely passive while the presenter does most of the talking, the workshop will be no more effective than when students sit passively in classes. The following tips for facilitators are taken from the workshop chapter of *Scientific Teaching* (Handelsman *et al.* 2007):

▶ Keep in mind the goals for the workshop.
▶ Identify the key points you want participants to learn from each section and then try to find ways to help participants discover them on their own.
▶ Reinforce the importance of collective problem-solving in finding multiple solutions for a single problem.

Instructions for Leading Activities

A. Welcome and Overview

The first step in leading any scientific teaching event is to create a welcoming environment that supports learning. If possible, let everyone introduce themselves and state why they are attending the workshop. State clearly the goals and intended outcomes for the workshop. As mentioned earlier, sharing the overview (or at least the intended learning outcomes) for the workshop with the participants will make the expectations clear and model good scientific teaching practices.

B. Topic 1: Backward Design

1. Pre-assessment brainstorm (large-group activity)
Begin by gauging the participants' attitudes toward assessment with the following question: "What is the most important function of assessment for you?" While participants call out answers, the presenter can record them on a white-

board, flip chart, or computer. This will show the presenter the variety of audience experience with using assessment for different reasons. The most typical response will likely be that participants use assessment to evaluate learning in order to assign course grades.

Facilitator tip: When soliciting responses from a large group, facilitators can collect responses from smaller groups rather than individuals to minimize the time required for the activity.

2. Think-pair-share (T-P-S): intended learning outcomes
Use the following activities to convey the importance of using outcomes to inform the design of learning activities and assessments.

  a. View and discuss the video: Show the opening sequence of *A Private Universe*, a documentary about the effects of misconceptions on learning, to illustrate the failure of traditional teaching methods to unseat common misconceptions about fundamental concepts in physics (Schneps *et al.* 1989; http://www.learner.org/resources/series28.html).

  After viewing the opening scene, prompt discussion with the question, "What is this example telling us about the effectiveness of traditional teaching methods?" After asking this question, give participants a few minutes to discuss it in small groups and then facilitate open discussion among the whole group. The goal of this discussion is to help participants appreciate the need for approaches like backward design that use specific learning outcomes to direct instructional design.

  Alternative: For similar videos focusing on topics in biology, see either *Minds of Our Own* (Schneps *et al.* 1997; www.learner.org/resources/series26.html) or *A Tiny World* (http://www.youtube.com/watch?v=0s7kc7Us4Hc).

  b. Think-Pair-Share (T-P-S): Following discussion of the video, prompt a T-P-S by asking, "What would you be most embarrassed to find that your students didn't know or couldn't do by the end of your course?" During a think-pair-share (see Table 4.1), participants first consider a problem or question individually, then discuss it with a partner or small group before sharing responses with the whole group. During the large group share, compile the list of embarrassing unmet learning outcomes on a whiteboard, flip chart, or computer. Facilitate a large group discussion of the list using the question, "How does the list generated by this activity compare to what you actually teach and what students actually do in class?" This discussion is intended to help participants contrast what they intend for their students to learn with how they teach them.

3. Mini-lecture: backward design

Use information from Chapter 1 and Figure 1.1 to convey the main point of backward design: the targeting of learning and assessment activities to intended learning outcomes helps drive meaningful learning. Introduce the four steps of backward design using the following questions.

▶ To identify intended learning outcomes, answer the question, "What should students know or be able to do by the end of the course?"

▶ To determine acceptable evidence of achievement, answer the question, "How will the instructor know if students have achieved the intended learning outcomes?"

▶ To plan learning experiences that promote meaningful learning, answer the question, "What instructional approaches will maximize the likelihood that students will achieve the intended learning outcomes?"

▶ To evaluate alignment of instruction materials, answer the question, "Does the assessment match the instruction?"

Provide an example of alignment for a specific topic (see Appendix A in Chapter 1 for examples).

C. Topic 2: Summative Assessment

1. Mini-lecture: summative assessment and alignment

Using information from Chapter 3, introduce summative assessments as the tools to provide evidence of whether students have achieved the intended learning outcomes. Since they are tied to course grades, summative assessments influence what and how students study. "The Montillation of Traxoline" (attributed to Judy Lanier, Professor, Dean Emeritus, Education Department, Michigan State University) is a nonsensical passage used to illustrate how little students need to understand in order to answer simple recall questions (https://www.sarc .miami.edu/ReinventionCenter/Public/Conference/2006Conference/assets/ DesigningPresentation.pdf). Ask one participant to read the passage aloud and then have participants call out answers to the associated quiz questions. After participants demonstrate that they can answer the quiz questions, prompt a large group discussion using the following questions:

▶ "Could students answer the quiz questions without understanding the passage?"

▶ "How does the cognitive level of the questions influence what and how students study?"

▶ "How do you think this passage compares to what students perceive when they read an introductory college science textbook for the first time?"

The goal of this discussion is to help participants realize that if assessment questions focus solely on recall, then students can perform well on assessments and still not learn what the instructor intends.

2. T-P-S: "Blooming" summative assessment questions

a. Mini-lecture: Begin with a mini-lecture to introduce participants to Bloom's Taxonomy (Table 2.1) and use Chapter 2 as a source for basic information on the taxonomy. Using Figure 2.1, highlight the differences between lower-order cognitive (LOC, Bloom levels 1 and 2) and higher-order cognitive (HOC, Bloom levels 4, 5, and 6) skill levels on the modified taxonomy. (Note that "application" can be categorized as either LOC or HOC, depending on the context within which the content was taught. If the question requires students to apply their understanding to a novel situation, then it is considered HOC; otherwise it is LOC.)

b. T-P-S: This T-P-S will provide participants with practice assigning cognitive Bloom levels to assessment questions so they can ultimately use Bloom's Taxonomy to evaluate alignment of their instructional materials. Using their own exams and Box 2.1, direct participants to:

▷ "Work with a neighbor to identify LOC and HOC questions on both your exams."

▷ "Identify an HOC question and state the intended learning outcome from your course that it addressed. (If you don't have an HOC question on the exam, use an LOC question.)"

▷ "Describe the in-class activities you used to prepare your students to achieve that intended learning outcome."

Once pairs have finished the previous activity, prompt a large group discussion of the results using the following questions:

> ▷ "Did you find alignment between intended learning outcomes, assessments, and learning activities?"
> ▷ "If not, where was the mismatch?"
> ▷ "How can you remedy the mismatch?"

Facilitator tip: During the large group discussion, help participants reflect on how following the steps of backward design can address misalignment issues in their instruction by prompting them to 1) state intended learning outcomes, 2) design assessments to gauge achievement of intended learning outcomes, 3) craft learning activities to aid achievement of intended learning outcomes, and 4) evaluate alignment.

D. Topic 3: Formative Assessment

1. Mini-lecture: formative assessment

Introduce formative assessment by revisiting backward design using Figure 1.1 and refocusing participants on step 3 in backward design (plan learning experiences that promote achievement). Use Figure 4.1 to illustrate the inextricable link between active learning and formative assessment. Finally, highlight the key aspects of good formative assessment: it engages students mentally in tasks that require reliance on previous knowledge, provides timely feedback, and aligns with intended learning outcomes.

2. T-P-S: gauging learning

Following the mini-lecture to introduce formative assessment, use the following questions to help participants articulate the importance of feedback from formative assessments in helping students gauge their progress during learning. Ask participants to first consider the questions individually and then discuss them with their neighbors or small groups. Finally, ask groups to report their answers to the large group for discussion. Answers can be compiled on a whiteboard, flip chart, or computer.

▶ "How do you know when you know something?"
▶ "How do you know when your students know something?"
▶ "How do your students know when they know something?"

Following compilation of the answers, prompt group discussion by asking participants to compare their responses to the three questions: "Are the responses the same or different? Why?"

Transition from this activity to the following one by asking participants to briefly share their ideas about what can be done in the classroom to help students know when they know something.

3. Formative assessment in action

Using information and methods from Chapter 4, model examples of how formative assessment can be used to:

▶ Engage students with questions, course content, and the behaviors and ways of thinking that represent science.
▶ Engage students' previous knowledge and help them construct new knowledge in a manner consistent with science.
▶ Provide timely feedback to both instructor and student about student learning.

Facilitator tip: The facilitator should take on the role of the classroom "teacher" and the participants should take on the role of "students" during these exercises, to afford the participants a more authentic experience of the engagement and feedback provided by formative assessment activities. If you use examples of successful activities from your own courses, you can provide participants with tips on implementation or insights into how students respond to a particular activity.

Using the following questions, facilitate a group discussion to help participants compare what they just experienced as "students" with what actual students experience in a traditional lecture classroom:

▶ "What did you learn or experience in the role of the student?"
▶ "What feedback did the assessments offer the students?"
▶ "How does this feedback compare to the feedback offered in a traditional lecture classroom?"

Facilitator tip: Some faculty may be uncomfortable in the role of student, especially if they exhibit the same misconceptions that students do. To help diffuse the situation, allow participants to share their negative experiences and solicit tips from other participants about how to deal with students who might also experience negative aspects of group learning activities.

E. Summary

Restate the key points of the workshop:

▶ Backward design is an approach to instructional design that focuses on outcomes.

▶ Aligning assessment and instruction with the intended learning outcomes facilitates meaningful learning.

▶ Summative assessment drives learning because it is used to determine grades.

▶ Formative assessment drives learning because it provides immediate feedback to help students and instructors gauge learning during the learning process, thereby affording students the opportunity to recalibrate their approach to studying.

F. Exit Assessment

Ask participants to write a one-minute response to each of the following prompts:

1. "What is the most important function of assessment? Is your answer different now than it was at the beginning of the workshop? If so, how and why?"
2. "How can different forms of assessment be used to drive student learning?"

Collect the responses. Compare the responses to question 1 with participant responses to the initial brainstorming activity to determine if participants changed their minds about the most important function of assessment. Use responses to question 2 to determine if participants understand formative assessment based on whether they can effectively compare and contrast it with summative assessment. Refer to Table 1.1 to evaluate responses.

Summary

When undertaking course reform, it is helpful to have a community of like-minded individuals for sharing ideas and lending support. This chapter provides the means for building that community by disseminating strategies for teaching reform based on best practices. The workshop can be used to train interested colleagues in the uses of assessment to improve student learning in college courses.

Glossary of Pedagogical Terms

Action Research: When faculty apply their scientific skills to answering questions about student learning or teaching efficacy.

Active learning: Instructional method employing learning activities that mentally engage students in a task such as answering a question, solving a problem or explaining a concept.

Alignment table for a course: Classifies each intended learning outcome, summative assessment question, and class activity according to Bloom's Taxonomy to determine whether these elements are aligned and whether various Bloom levels are appropriately represented across the design of a course.

Alternative Form Reliability: A student will have relatively the same result when they take alternative (isomorphic) forms of the same test at two different times because the different items assess the same content or skills at the same cognitive level.

Backward design: A goal-oriented approach to instructional design that focuses on what students will learn rather than what instructors will teach. This approach employs a three-step process that begins with identification of intended learning outcomes, followed by design of the assessments to evaluate achievement, and ending with the learning activities to help students achieve the intended learning outcomes.

Bloom's Taxonomy of the Cognitive Domain: A classification of educational goals for the development of a student's intellectual skills. The original taxonomy included knowledge, comprehension, application, analysis, synthesis, and evaluation. In recent years the taxonomy has been modified to turn the nouns into verbs and switch the two higher levels: recall, understand, apply, analyze, evaluate, and synthesize.

Brainstorming: A low-pressure learning activity in which students list everything they already know on a given topic or in response to a simple, open-ended question.

Case studies: Learning activities that present students with a scenario involving a problem, dilemma, or situation that must be solved, resolved, or evaluated by students.

Classical test theory: A statistical framework that relies on test-score models, where the test scores for a given test group are analyzed and items are modified on the basis of the group results. There is great emphasis on finding and using the most suitable group for test development.

Clickers: Personal immediate-response devices that students use to answer in-class questions.

Collaborative learning: Active learning activities in which students work together in a small group and receive a group grade.

Cooperative learning: A form of collaborative learning in which students work together in small groups, but are graded individually.

Concept inventory: A test designed to assess the major core concepts of a discipline. Concept inventories are as jargon-free as possible so that they can be used for both pre- and post-testing. These assessments have been validated by experts in the field and their reliability established. Most concept inventories are in a multiple-choice format.

Construct Validity: The *item* measures what it is intended to measure (requires affirmation by experts).

Constructivism: The view that students must build their own understanding of a topic. It contrasts with the transmissionist view of learning, which implies that instructors can transmit conceptual understanding to students simply by telling them about it.

Content Validity: The *test* presents the key concepts and misconceptions for a given domain as affirmed by experts. Content validity increases as the number of relevant items on the test increases.

Cross-over research format: A testing format in which students are divided into groups A and B. For a particular teaching intervention, group A serves as the treatment group while group B serves as the control group. The groups are then switched for a different topic that uses the same pedagogical approach. Student learning is assessed with pre- and post-tests, perception surveys, interviews, or another method to determine if the new approach was effective for improving learning.

Diagnostic question cluster: Groups of related questions intended to reveal student reasoning and potential misconceptions.

Discipline-based education research: When evidence generated from action research moves into the realm of education research and publication.

Effect size: A measure of differences between groups that emphasizes the magnitude, rather than the statistical significance, of the difference.

Exam construction guide: A table in which the instructor lists the percentages of course time devoted to each of the general topics covered, and subdivides these percentages to indicate the proportions that students spent working on either lower- or higher-order cognitive skills. The table serves as a guide for creating the appropriate number of exam questions at the corresponding cognitive levels.

"Engaugements": Challenging learning activities that both engage students and provide feedback to allow students to gauge their own learning progress.

Face Validity: The test has the appearance of measuring what it intends to measure, thereby motivating students to do their best. The assumption is that students performed to the best of their abilities on the test.

FLAG (Field-tested Learning Assessment Guide): A comprehensive online assessment website.

Formative assessment: Assessment *for* learning; activities that students engage in, either in or out of class, that give the student feedback on their current understanding and provide practice in mastering the intended skill or content associated with the class.

Grounded theory: A method for systematically analyzing data in order to develop a theory. It is in contrast to hypothesis-directed research and is similar to interpretive research.

Immediate feedback assessment technique (IF-AT): Also known as answer until correct (AUC); multiple-choice assessment in which students answer by scratching an opaque substance off of answer choices. Correct answers are designated by an asterisk. Students can continue to answer until they get the correct response, but the points earned diminish with each unsuccessful attempt.

Inquiry-based learning: Instructional method that uses questioning to drive acquisition of knowledge or understanding.

Interleaving: A method to structure learning and practice activities to maximize long-term retention of material. Interleaving (a,b,c,b,a,c,a,c,b) is in contrast to blocking (a,a,a,b,b,b,c,c,c).

Internal Consistency: Measures whether pair-wise items on a test that should measure the same content, or latent variable, actually do so.

Interpretive research: Collecting qualitative data for generating quantitative data, or mining qualitative data for discovery.

Inter-Rater Reliability: The consistency of scoring free-response items by different graders.

Intra-Rater Reliability: The consistency of scoring free-response items by the same grader at different times.

Isomorphic items: Items that are written to assess the same content. These are also called alternative forms.

Item: An individual question or task to which a student will respond.

Item Difficulty: The percentage of students answering the item correctly (a confusing but unfortunately established convention; note that the easiest items have the highest item difficulty).

Item Discrimination: The item's ability to discriminate between students of different academic ability (those who have mastered the material and those who haven't). The students with mastery of the content are more likely to answer an item correctly.

Item Response Theory: A statistical framework that focuses on the analysis of each item in relation to an examinee's ability and is considered to be test-independent.

Just-in-Time Teaching (JiTT): Instructional strategy in which instructors use student responses to a pre-class assessment of students' prior knowledge to guide decisions about instruction.

Learning gains: Data from both identical or isomorphic pre-and post-tests can be used to calculate learning gains. The normalized gain for a single student <g> is calculated as the percent gain divided by the maximum possible gain. It can also be calculated as the average gain for the class <g_{class}>.

Learning goals: Broad, desired impacts on student learning; typically stated with verbs like understand, know, or appreciate, which are not easily assessable.

Learning objectives: Originally defined as the instructor's goals for a course, or the teaching objectives. Now often used as a synonym for intended learning outcomes, since "learning" objectives implies a focus on students.

Intended learning outcomes: What students should be able to do by the end of a course. This term clearly focuses on student learning rather than the instructor's teaching plan.

Metacognition: Understanding one's own learning.

Model-based learning (MBL): A cooperative learning strategy in which groups of students work to represent and comprehend complex systems and phenomena with models.

One-minute paper: A learning activity that involves students writing for approximately one minute on a topic or in response to an open-ended question.

Peer-led team learning: A cooperative learning strategy in which veteran students act as learning assistants to mentor and facilitate teams of students.

Pre- and post-test format: When an instructor uses an assessment before and after teaching content or skills that the assessment measures. This format allows the instructor to assess learning gains of individual students or all students in the class.

Problem-based learning (PBL): A cooperative learning strategy in which students work in groups to solve complex problems.

Process-oriented guided inquiry learning (POGIL): A highly structured cooperative learning method in which students are assigned specific roles within small groups so that all are actively engaged in the process of guided inquiry learning.

Reading assessment: A collaborative learning activity requiring students to work as a group to design an activity that will assess comprehension of a reading assignment.

Reliability: Addresses the consistency of a set of measures by a given assessment and is based on whether the test results are repeatable and not subject to random error.

Scientific teaching: A pedagogical approach that reflects the rigorous methodology with which scientists approach their disciplinary research. This approach is based on constructivism and promotes inclusive use of active learning and assessment.

Sequential course format: An assessment method in which instructors use pre- and post-testing to compare different groups of students in the same course over subsequent semesters. This allows an instructor to determine whether a particular teaching intervention can consistently elicit the same level of student achievement with different groups of students.

Spacing effect: The research finding that distributed studying across time is more effective for long-term retention of material than massed studying in a single short period (cramming). The greater the time between study sessions, the longer the retention.

STEM: Science, Technology, Engineering, and Mathematics.

Strip sequence: A learning activity that requires students to put the jumbled steps of a process in the appropriate order.

Summative assessments: Exams, written or oral reports, projects, etc. administered or assigned at the end of a unit or course as the assessment *of* learning. These are generally high-stakes assessments that heavily influence student grades.

Team-based learning: A cooperative learning strategy in which students work in teams to accomplish goals and construct understanding through structured tasks.

Testing effect: The research finding that students learn better from repeated testing than from repeated study of course material. Testing not only enhances learning but slows the rate of forgetting.

Test–Re-Test Reliability: A student or group of students will score similarly if they take the same test at two different times without intervention.

Think-Pair-Share: A learning activity that involves students first pondering a question individually and then discussing the topic with a partner or group before sharing their answer with the entire class.

Validated Instruments: Assessments that measure what they are supposed to measure, and do so in such a manner as to consistently distinguish between individuals with disparate abilities.

Validity: Addresses whether the test measures what it is supposed to measure.

References

Adams, W., S. Reid, R. LeMaster, S. McKagan, K. Perkins, M. Dubson, and C. Wieman. 2008a. A study of educational simulations, Part I: Engagement and learning. *Journal of Interactive Learning Research* 19: 397–419.

Adams, W., S. Reid, R. LeMaster, S. McKagan, K. Perkins, M. Dubson, and C. Wieman. 2008b. A study of educational simulations, Part II: Interface design. *Journal of Interactive Learning Research* 19: 551–577.

Albanese, M. 1993. Type K and other complex multiple-choice items: An analysis of research and item properties. *Educational Measurement: Issues and Practice* 12: 28–33.

Allan, J. 1996. Learning outcomes in higher education. *Studies in Higher Education* 21: 93–105.

Allen, D. 1997. Bringing problem-based learning to the introductory biology classroom. In *Student Active Science*. A. McNeal and C. D'Avanzo, eds. New York: Saunders.

Allen, D. and K. Tanner. 2002. Approaches in cell biology teaching. *CBE—Life Sciences Education* 1:3–5.

Allen, D. and K. Tanner. 2003. Learning content in context: Problem-based learning. *CBE—Life Sciences Education* 2: 73–81.

Allen, D. and K. Tanner. 2006. Rubrics: Tools for making learning goals and evaluation criteria explicit for both teachers and learners. *CBE—Life Sciences Education* 5: 197–203.

Allen, K. 2006. The Statistics Concept Inventory: Development and analysis of a cognitive assessment instrument in statistics (Doctoral dissertation). Available from ProQuest Dissertations and Theses database (UMI No. 3212015).

Alias, M. 2005. Assessment of learning outcomes: Validity and reliability of classroom tests. *World Transactions on Engineering and Technology Education* 4: 235–238.

American Association for the Advancement of Science (AAAS). 2007. *Atlas of Science Literacy, Volume 2.* Washington, DC: AAAS Press.

American Association for the Advancement of Science (AAAS). 2011. *Vision and Change in Undergraduate Biology Education: A Call to Action.* Washington, DC: AAAS Press.

Anderson, T. 2007. Bridging the educational research-teaching practice gap. The importance of bridging the gap between science education research and its application in biochemistry teaching and learning: Barriers and strategies. *Biochemistry and Molecular Biology Education* 35: 465–470.

Anderson, D., K. Fisher, and G. Norman. 2002. Development and evaluation of the conceptual inventory of natural selection. *Journal of Research in Science Teaching* 39: 952–978.

Anderson, L., D. Krathwohl, and B. Bloom. 2001. *A Taxonomy for Learning, Teaching, and Assessing: A Revision of Bloom's Taxonomy of Educational Objectives*. New York: Longman.

Anderson, L. and L. Sosniak. 1994. *Bloom's Taxonomy of Educational Objectives: A Forty-Year Retrospective*. Chicago: The National Society for the Study of Education. Distributed by the University of Chicago Press.

Angelo, T. 1995. Reassessing (and defining) assessment. *AAHE Bulletin* 48: 7–9.

Angelo, T. and K. Cross. 1993. *Classroom Assessment Techniques: A Handbook for College Teachers*. San Francisco: Jossey-Bass Publishers.

Association of American Medical Colleges and Howard Hughes Medical Institute. 2009. *Scientific Foundations for Future Physicians*. Retrieved from http://www.aamc.org/scientificfoundations

Ausubel, D. 2000. *The Acquisition and Retention of Knowledge: A Cognitive View*. Boston: Kluwer Academic Publishers.

Baddeley, A., V. Lewis, M. Eldridge, and N. Thomson. 1984. Attention and retrieval from long-term memory. *Journal of Experimental Psychology: General* 113: 518–540.

Baker, F. 2001. *The Basics of Item Response Theory*. 2nd ed. College Park, MD: ERIC Clearinghouse on Assessment and Evaluation.

Balch, W. 1998. Practice versus review exams and final exam performance. *Teaching of Psychology* 25: 181–185.

Bardar, E., E. Prather, K. Brecher, and T. Slater. 2007. Development and validation of the Light and Spectroscopy Concept Inventory. *Astronomy Education Review* 5: 103–113.

Barkley, E., K. Cross, and C. Major. 2005. *Collaborative Learning Techniques: A Handbook for College Faculty*. San Francisco: Josey-Bass.

Barr, R. and J. Tagg. 1995. From teaching to learning: a new paradigm for undergraduate education. *Change* 27: 12–25.

Beichner, R. and J. Saul. 2003. Introduction to the SCALE-UP (Student-Centered Activities for Large Enrollment Undergraduate Programs) project. *Proceedings of the International School of Physics*. Retrieved from http://www.ncsu.edu/per/Articles/Varenna_SCALEUP_Paper.pdf

Biggs, J. 1979. Individual differences in study processes and the quality of learning outcomes. *Higher Education* 8: 381–394.

Biggs, J. 1993. What do inventories of students' learning processes really measure? A theoretical review and clarification. *British Journal of Educational Psychology* 63: 3–19.

Biggs, J. and K. Collins. 1982. *Evaluating the Quality of Learning: The SOLO Taxonomy*. New York: Academic Press.

Biggs, J., D. Kember, and D. Leung. 2001. The revised two-factor study process questionnaire: R-SPQ-2F. *British Journal of Educational Psychology* 71: 133–149.

Biggs, J. and C. Tang. 2009. *Teaching for Quality Learning at University: What the Student Does.* Maidenhead, UK: McGraw-Hill.

Biron, A., C. Loiselle, and M. Lavoie-Tremblay. 2009. Work interruptions and their contribution to medication administration errors: An evidence review. *Worldviews on Evidence-Based Nursing* 6: 70–86.

Bjork, R. 1979. Information-processing analysis of college teaching. *Educational Psychologist* 14: 15–23.

Bjork, R. 1994a. Memory and metamemory considerations in the training of human beings. In *Metacognition: Knowing about Knowing.* J. Metcalfe and A. Shimamura, eds. Cambridge, MA: MIT Press.

Bjork, R. 1994b. Institutional impediments to effective training. In *Learning, Remembering, Believing: Enhancing Human Performance.* D. Druckman and R. Bjork, eds. Washington, DC: National Academies Press.

Black, P. 1998. Formative assessment: Raising standards inside the classroom. *School Science Review* 80: 39–46.

Black, P. and D. Wiliam. 1998. Inside the black box: Raising standards through classroom assessment. *Phi Delta Kappa* 80: 139–148.

Black, P. and D. Wiliam. 2004. The formative purpose: Assessment must first promote learning. *Yearbook of the National Society for the Study of Education* 103: 20–50.

Bloom, B. 1956. *Taxonomy of Educational Objectives: The Classification of Educational Goals.* 1st ed. New York: D. McKay.

Bloom, B., J. Hastings, and G. Madaus. 1971. *Handbook on Formative and Summative Evaluation of Student Learning.* New York: McGraw-Hill.

Boehrer, J. and M. Linsky. 1990. Teaching with cases: Learning to question. *New Directions for Teaching and Learning* 42: 41–57.

Bowling, B., E. Acra, L. Wang, M. Myers, G. Dean, G. Markle, . . . C. Huether. 2008. Development and evaluation of a genetics literacy assessment instrument for undergraduates. *Genetics* 178: 15–22.

Bransford, J., A. Brown, and R. Cocking. 1999. *How People Learn: Brain, Mind, Experience, and School.* Washington, DC: National Academies Press.

Bresciani M., C. Zelna, and J. Anderson. 2004. *Assessing Student Learning and Development: A Handbook for Practitioners.* Washington, DC: National Association of Student Personnel Administrators.

Brewer, C. 2004. Near real-time assessment of student learning and understanding in biology courses. *Bioscience* 54: 1034–1039.

Broadfoot, P. and P. Black. 2004. Redefining assessment? The first ten years of assessment in education. *Assessment in Education Principles Policy and Practice* 11: 7–26.

Brown, P. 2010. Process-oriented guided-inquiry learning in an introductory anatomy and physiology course with a diverse student population. *Advances in Physiology Education* 34: 150–155.

Buckley, B. 2000. Interactive multi-media and model-based learning in biology. *International Journal of Science Education* 22: 895–935.

Bunce, D., J. VandenPlas, K. Neiles, and E. Flens. 2010. Development of a valid and reliable student-achievement and process-skills instrument. *Journal of College Science Teaching* 39: 50–55.

Butler, A., J. Karpicke, and H. Roediger. 2007. The effect of type and timing of feedback on learning from multiple-choice tests. *Journal of Experimental Psychology: Applied* 13: 273–281.

Caldwell, J. 2007. Clickers in the large classroom: Current research and best-practice tips. *CBE—Life Sciences Education* 6: 9–20.

Carmichael, J. 2009. Team-based learning enhances performance in introductory biology. *Journal of College Science Teaching* 38: 54–61.

Carmines, E. and R. Zeller. 1979. *Reliability and Validity Assessment.* Beverly Hills, CA: Sage.

Carter, C. 1998. Assessment: Shifting the responsibility. *Journal of Secondary Gifted Education* 9: 68–75.

Chi, M., N. de Leeuw, M. Chiu, and C. LaVancher. 1994. Eliciting self-explanations improves understanding. *Cognitive Science* 18: 439–477.

Chun, M. and N. Turk-Browne. 2007. Interactions between attention and memory. *Current Opinion in Neurobiology* 17: 177–184.

Cohen, J. 1969. *Statistical Power Analysis for the Behavioral Sciences.* New York: Academic Press.

Cohen, D. and J. Henle. 1995. The pyramid exam. *UME Trends* July: 2.

Cohen, L. and L. Manion. 1977. *A Guide to Teaching Practice.* London: Routledge.

Coil, D., M. Wenderoth, M. Cunningham, and C. Dirks. 2010. Teaching the process of science: Faculty perceptions and an effective methodology. *CBE—Life Sciences Education* 9: 524–535.

Coleman, E., A. Brown, and I. Rivkin. 1997. The effect of instructional explanations on learning from scientific texts. *Journal of the Learning Sciences* 6: 347–365.

College Board. 2009. *College Board Standards for College Success.* Retrieved from http://apcentral.collegeboard.com/apc/public/repository/cbscs-science-standards-2009.pdf.

Cook, D. and T. Beckman. 2006. Current concepts in validity and reliability for psychometric Instruments: Theory and application. *The American Journal of Medicine* 119: 166.e7–166.e16.

Corbin, J. and A. Strauss. 1990. Grounded theory research: Procedures, canons, and evaluative criteria. *Qualitative Sociology* 13: 3–21.

Cotner, S., B. Fall, S. Wick, J. Walker, and P. Baepler. 2008. Rapid feedback assessment methods: Can we improve engagement and preparation for exams in large enrollment courses? *Journal of Science Education and Technology* 15: 437–443.

Craik, F., R. Govoni, M. Naveh-Benjamin, and N. Anderson. 1996. The effects of divided attention on encoding and retrieval processes in human memory. *Journal of Experimental Psychology: General* 125: 159–180.

Cronbach, L. 1951. Coefficient alpha and the internal structure of tests. *Psychometrika* 16: 297–334.

Crowe, A., C. Dirks, and M. Wenderoth. 2008. Biology in bloom: Implementing Bloom's Taxonomy to enhance student learning in biology. *CBE—Life Sciences Education* 7: 368–381.

Crowell, B. 2000. *Newtonian Physics*. Fullerton, CA: Light and Matter.

D'Avanzo, C. 2008. Biology concept inventories: Overview, status, and next steps. *Bioscience* 58: 1079–1085.

Dewey, J. 1966. *Democracy and Education: An Introduction to the Philosophy of Education*. New York: The Free Press.

de Koning, B., H. Tabbers, R. Rikers, and F. Paas. 2010. Attention guidance in learning from a complex animation: Seeing is understanding? *Learning and Instruction* 20: 111–122.

deWinstanley, P. and R. Bjork. 2002. Successful lecturing: Presenting information in ways that engage effective processing. *New Directions for Teaching and Learning* 2002:19–31.

Dickie, L. 2003. Approach to learning, the cognitive demands of assessment, and achievement in physics. *Canadian Journal of Higher Education* 33: 87–111.

Ding, L., R. Chabay, B. Sherwood, and R. Beichner. 2006. Evaluating an electricity and magnetism assessment tool. Brief electricity and magnetism assessment. *Physical Review Special Topics—Physics Education Research* 2:1–7.

Dweck, C. and D. Schunk. 2000. Self-theories: Their role in motivation, personality, and development. *Contemporary Psychology* 45: 554.

Eberlein, T., J. Kampmeier, V. Minderhout, R. Moog, T. Platt, P. Varma-Nelson, and H. White. 2008. Pedagogies of engagement in science: A comparison of PBL, POGIL, and PLTL. *Biochemistry and Molecular Biology Education* 36: 262–273.

Ebert-May, D., J. Batzli, and H. Lim. 2003. Disciplinary research strategies for assessment of learning. *BioScience* 53: 1221–1228.

Educational Testing Service (ETS). 2012. *Graduate Record Examinations® Physics Test Practice Book*. Retrieved from http://www.ets.org/gre/subject/about/content/physics

Eisner, E. 1979. *The Educational Imagination*. New York: Macmillan.

Engelhardt, P. 2009. An introduction to classical test theory as applied to conceptual multiple-choice tests. In *Getting Started in Physics Education Research*. C. Henderson and K. Harper, eds. *Reviews in Physics Education Research* 2. Retrieved from http://www.per-central.org/items/detail.cfm?ID=8807

Entwistle, N. and P. Ramsden. 1982. *Understanding Student Learning*. New York: Nichols Publishing.

Epstein, J. 2005. Development and validation of the Calculus Concept Inventory. Retrieved from http://math.unipa.it/~grim/21_project/21_charlotte_EpsteinPaperEdit.pdf

Epstein, M., B. Epstein, and G. Brosvic. 2001. Immediate feedback during academic testing. *Psychological Reports* 88: 889–894.

Epstein, M., A. Lazarus, T. Calvano, K. Matthews, R. Hendel, B. Epstein, and G. Brosvic. 2002. Immediate feedback assessment technique promotes learning and corrects inaccurate first responses. *Psychological Record* 52: 187–201.

Erickson, F. 1986. Qualitative research on teaching. In *Handbook for Research on Teaching*. M. Whittrock, ed. New York: Macmillian.

Ericsson, K., R. Krampe, and C. Tesch-Römer. 1993. The role of deliberate practice in the acquisition of expert performance. *Psychological Review* 100: 363–406.

Farrell, J., R. Moog, and J. Spencer. 1999. A guided inquiry chemistry course. *Journal of Chemical Education* 76: 570–574.

Fazio, L., P. Agarwal, E. Marsh, and H. Roediger. 2010. Memorial consequences of multiple-choice testing on immediate and delayed tests. *Memory and Cognition* 38: 407–418.

Flynn, E., K. Barker, J. Gibson, R. Pearson, B. Berger, and L. Smith. 1999. Impact of disruptions and distractions on dispensing errors in an ambulatory care pharmacy. *American Journal of Health System Pharmacy* 56: 1319–1325.

Foster, D. and A. Poppers. 2009. Using formative assessment to drive learning. *The Silicon Valley Mathematics Initiative: A Twelve-Year Research and Development Project*. Morgan Hill, CA: Silicon Valley Mathematics Initiative.

Freeman, S. 2010. *Biological Science*. 4th ed. San Francisco: Benjamin Cummings.

Freeman, S., E. O'Connor, J. Parks, M. Cunningham, D. Hurley, D. Haak, C. Dirks, and M. Wenderoth. 2007. Prescribed active learning increases performance in introductory biology. *CBE—Life Sciences Education* 6: 132–139.

Freeman, S. and J. Parks. 2010. How accurate is peer grading? *CBE—Life Sciences Education* 9: 482–488.

Friedman, H. 1987. Immediate feedback, no return test procedure for introductory courses. *Teaching of Psychology* 14: 244.

Frisbie, D. 1990. "The evolution of the multiple true-false item format." Paper presented at the Annual Meeting of the National Council on Measurement in Education. Boston, MA.

Froyd, J. 2008. Evidence for the efficacy of student-active learning pedagogies. *Project Kaleidoscope*. Retrieved from http://www.pkal.org/documents/BibliographyofSALPedagogies.cfm

Gafney, L. and P. Varma-Nelson. 2008. *Peer-Led Team Learning: Evaluation, Dissemination, and Institutionalization of a College Level Initiative*. Dordrecht, The Netherlands: Springer.

Gibbs, G. 1992. *Improving the Quality of Student Learning*. Bristol, UK: Technical and Educational Services.

Gilbert, J. and C. Boulter. 1998. Learning science through models and modeling. In *The International Handbook of Science Education*. B. Fraser and K. Tobin, eds. Dordrecht, The Netherlands: Kluwer.

Glover, J. 1989. The "testing" phenomenon: Not gone but nearly forgotten. *Journal of Educational Psychology* 81: 392–399.

Gosser, D., M. Cracolice, J. Kampmeier, V. Roth, V. Strozak, and P. Varma-Nelson. 2001. *Peer-Led Team Learning: A Guidebook.* Upper Saddle River, NJ: Prentice Hall.

Green, P. 2002. *Peer Instruction in Astronomy.* Boston: Addison-Wesley.

Guthrie, R. and A. Carlin. 2004. Waking the dead: Using interactive technology to engage passive listeners in the classroom. *Proceedings of the 10th Americas Conference on Information Systems.* New York, NY. Retrieved from http://www.mhhe.com/cps/docs/CPSWP_WakindDead082003.pdf

Ha, M., R. Nehm, M. Urban-Lurain, and J. Merrill. 2011. Applying computerized-scoring models of written biological explanations across courses and colleges: Prospects and limitations. *CBE—Life Sciences Education* 10: 379–393.

Hake, R. 2001. Suggestions for administering and reporting pre/post diagnostic tests. Retrieved from http://physics.indiana.edu/~hake/

Haladyna, T., S. Downing, and M. Rodriguez. 2002. A review of multiple-choice item-writing guidelines for classroom assessment. *Applied Measurement in Education* 15: 309–334.

Hall, R., H. Collier, M. Thomas, and M. Hilgers. 2005. A student response system for increasing student engagement, motivation, and learning in high enrollment chemistry lectures. *Proceedings of the 11th Americas Conference on Information Systems.* Omaha, NE. Retrieved from http://lite.msu.edu/documents/hall_et_al_srs_amis_procee

Halloun, I. and D. Hestenes. 1998. Interpreting VASS dimensions and profiles. *Science and Education* 7: 553–577.

Hambleton, R. and R. Jones. 1993. Comparison of classical test theory and item response theory and their applications to test development. *Educational Measurement: Issues and Practice* 58: 357–381.

Handelsman, J., S. Miller, and C. Pfund. 2007. *Scientific Teaching.* New York: W.H. Freeman.

Harden, R.M. 2002. Learning outcomes and instructional objectives: Is there a difference? *Medical and Dental Education Medical Teacher* 24: 151–155.

Haring-Smith, T. 1993. *Learning Together: An Introduction to Collaborative Learning.* New York: Longman.

Hartberg, Y., A. Gunersel, N. Simpson, and V. Balester. 2008. Development of student writing in biochemistry using calibrated peer review. *Journal of the Scholarship of Teaching and Learning* 2: 29–44.

Hartley, L., B. Wilke, J. Schramm, C. D'Avanzo, and C. Anderson. 2011. College students' understanding of the carbon cycle: Contrasting principle-based and informal reasoning. *Bioscience* 61: 65–75.

Herskovic, P. 1999. Reutilization of multiple-choice questions. *Medical Teacher* 21: 430–431.

Herreid, C. 2006. Clicker cases. *Journal of College Science Teaching* 36: 43–47.

Herreid, C. 2007. *Start with a Story.* Arlington, VA: National Science Teachers' Association Press.

Herreid, C. 2011. Case study teaching. *New Directions for Teaching and Learning* 128: 31–40.

Hestenes, D., M. Wells, and G. Swackhamer. 1992. Force Concept Inventory. *The Physics Teacher* 30: 141–158.

Hufnagel, B. 2002. Development of the Astronomy Diagnostic Test. *The Astronomy Education Review* 1: 47–51.

Jacobi, A., J. Martin, J. Mitchell, and T. Newell. 2003. A concept inventory for heat transfer. *Proceedings of the Frontiers in Education Conference, Boulder, CO.*

Jacoby, L. 1983. Remembering the data: Analyzing interactive processes in reading. *Journal of Verbal Learning and Verbal Behavior* 22: 485–508.

Jacoby, L., T. Jones, and P. Dolan. 1998. Two effects of repetition: Support for a dual-process model of knowledge judgments and exclusion errors. *Psychonomic Bulletin and Review* 5: 705–709.

Jensen, J. and A. Lawson. 2011. Effects of collaborative group composition and inquiry instruction on reasoning gains and achievement in undergraduate biology. *CBE—Life Sciences Education* 10: 64–73.

Johnson, D., R. Johnson, and E. Holubec. 1990. *Circles of Learning: Cooperation in the Classroom.* Edina, MN: Interaction Book Company.

Johnson, D., R. Johnson, and K. Smith. 1998. *Active Learning: Cooperation in the College Classroom.* 2nd ed. Edina, MN: Interaction Book Company.

Jonassen, D., K. Beissner, and M. Yacci. 1993. *Structural Knowledge: Techniques for Representing, Conveying, and Acquiring Structural Knowledge.* Hillsdale, NJ: Lawrence Erlbaum Associates, Inc.

Kagan, S. 1994. *Cooperative Learning.* San Juan Capistrano, CA: Cooperative Learning.

Karpicke, J. and J. Blunt. 2011. Retrieval practice produces more learning than elaborative studying with concept mapping. *Science* 331: 772–775.

Karpicke, J. and H. Roediger. 2007. Repeated retrieval during learning is the key to long-term retention. *Journal of Memory and Language* 57: 151–162.

Karpicke, J. and H. Roediger. 2008. The critical importance of retrieval for learning. *Science* 319: 966–968.

Keefer, K. 1969. Self-prediction of academic achievement by college students. *Journal of Educational Research* 63: 53–56.

Kember, D. and D. Leung. 2008. Establishing the validity and reliability of course evaluation questionnaires. *Assessment and Evaluation in Higher Education* 33: 341–353.

Kember, D. and C. McNaught. 2007. *Enhancing University Teaching: Lessons from Research into Award-Winning Teachers.* Abingdon, UK: Routledge.

Klionsky, D. 2004. Talking biology: Teaching outside the book—and the lecture. *CBE—Life Sciences Education* 3: 204–211.

Klymkowsky, M., S. Underwood, and R. Garvin-Doxas. 2010. "The Biological Concepts Instrument (BCI): A diagnostic tool to reveal student thinking." Retrieved from http://arxiv.org/ftp/arxiv/papers/1012/1012.4501.pdf

Koriat, A. and R. Bjork. 2005. Illusions of competence in monitoring one's knowledge during study. *Journal of Experimental Psychology: Learning, Memory, and Cognition* 31: 187–194.

Kornell, N. and R. Bjork. 2007. The promise and perils of self-regulated study. *Psychonomic Bulletin and Review* 14: 219–224.

Kornell, N., M. Hays, and R. Bjork. 2009. Unsuccessful retrieval attempts enhance subsequent learning. *Journal of Experimental Psychology: Learning, Memory, and Cognition* 35: 989–998.

Knight, J. and W. Wood. 2005. Teaching more by lecturing less. *CBE—Life Sciences Education* 4: 298–310.

Krathwohl, D. 2002. A revision of Bloom's Taxonomy: An overview. *Theory into Practice* 41: 212–218.

Krause, S., J. Decker, and R. Griffin. 2003. Using a materials concept inventory to assess conceptual gain in introductory materials engineering courses. *Proceedings of the Frontiers in Education Conference, Boulder, CO.*

Kuder, G. and M. Richardson. 1937. The theory of the estimation of test reliability. *Psychometrik,* 2: 151–160.

Landis, J. and G. Koch. 1977. The measurement of observer agreement for categorical data. *Biometrics* 33: 159–174.

Larsen, D., A. Butler, and H. Roediger. 2008. Test-enhanced learning in medical education. *Medical Education* 42: 959–966.

Larsen, D., A. Butler, and H. Roediger. 2009. Repeated testing improves long-term retention relative to repeated study: A randomised controlled trial. *Medical Education* 43: 1174–1181.

Lawson, A. 1978. The development and validation of a classroom test of formal reasoning. *Journal of Research in Science Teaching* 15: 11–24.

Levesque, S. 2011. Using clickers to facilitate development of problem-solving skills. *CBE—Life Sciences Education* 10: 406–417.

Libarkin, J. and S. Anderson. 2005. Assessment of learning in entry-level geoscience courses: Results from the Geoscience Concept Inventory. *Journal of Geoscience Education* 53: 394–401.

Lim, K., H. Lee, and B. Grabowski. 2009. Does concept-mapping strategy work for everyone? The levels of generativity and learners' self-regulated learning skills. *British Journal of Educational Technology* 40: 606–618.

Lindell, R. and J. Olsen. 2002. *Developing the Lunar Phases Concept Inventory.* Paper presented at the Physics Education Research Conference. Boise, ID.

Lopatto, D. 2004. Survey of Undergraduate Research Experiences (SURE): First Findings. *CBE—Life Sciences Education* 3: 270–277.

Lord, F. 1953. The relation of test score to the trait underlying the test. *Educational and Psychological Measurement* 13: 517–548.

Lord, T. and S. Baviskar. 2007. Moving sudents from information recitation to information understanding. *Journal of College Science Teaching* 36: 40–44.

Lovett, M. 2008. "Teaching Metacognition." Presentation to the Educause Learning Initiative Annual Meeting. Retrieved from http://net.educause.edu/upload/presentations/ELI081/FS03/Metacognition-ELI.pdf

Luft, J. 1999. Rubrics: Design and use in science teacher education. *Journal of Science Teacher Education* 10: 107–121.

Lunsford, E. and C. Melear. 2004. Using scoring rubrics to evaluate inquiry. *Journal of College Science Teaching* 34: 34–38.

MacDonald-Ross, M. 1973. Behavioural Objectives: A Critical Review. *Instructional Science* 2: 1–52.

Mager, R. 1962. *Preparing Instructional Objectives.* Belmont, CA: David Lake Publishers.

Maloney, D., T. O'Kuma, C. Hieggelke, and A. Heuvelen. 2001. Surveying students' conceptual knowledge of electricity and magnetism. *American Journal of Physics* 69: S12–S23.

Marbach-Ad, G., V. Briken, N. El-Sayed, K. Frauwirth, B. Fredericksen, S. Hutcheson, . . . A. Smith. 2009. Assessing student understanding of host pathogen interactions using a concept inventory. *Journal of Microbiology and Biology Education* 10: 43–50.

Marbach-Ad, G., K. McAdams, S. Benson, V. Briken, L. Cathcart, M. Chase, . . . A. Smith. 2010. A model for using a concept inventory as a tool for students' assessment and faculty professional development. *CBE—Life Sciences Education* 9: 408–416.

Marchese, T. 1992. Assessing Learning at Harvard. *American Association for Higher Education (AAHE) Bulletin* 44(6): 3–7.

Marsh, E., H. Roediger, R. Bjork, and E. Bjork. 2007. The memorial consequences of multiple-choice testing. *Psychonomic Bulletin and Review* 14: 194–199.

Martin-Morris, L. and R. Wright. 2007. Point recapture: Learning from exams. Retrieved from http://depts.washington.edu/sotl/2007/Martin-Morris.ppt

Marton, F., D. Hounsell, and N. Entwistle. 1984. *The Experience of Learning: Implications for Teaching and and Studying in Higher Education.* Edinburgh: Scottish Academic Press.

Marton, F. and R. Saljo. 1976. On qualitative differences in learning—II Outcome as a function of the learner's conception of the task. *British Journal of Educational Psychology* 46: 115–127.

Marx, J. and K. Cummings. 2007. Normalized change. *American Journal of Physics* 75: 87–91.

Marzano, R. and J. Kendall. 2008. *Designing and Assessing Educational Objectives: Applying the New Taxonomy.* Thousand Oaks, CA: Corwin Press.

Maskiewicz, A., H. Griscom, and N. Welch. 2012. Using targeted active learning exercises and diagnostic question clusters to improve students' understanding of carbon cycling in ecosystems. *CBE—Life Sciences Education* 11: 58–67.

Mayer, R., A. Stull, K. DeLeeuw, K. Almeroth, B. Bimber, D. Chun, M. Bulger, J. Campbell, A. Knight, and H. Zang. 2009. Clickers in college classrooms: Fostering learning with questioning methods in large lecture classes. *Contemporary Educational Psychology* 34: 51–57.

Mazur, E. 1997. *Peer Instruction: A User's Manual.* Upper Saddle River, NJ: Prentice Hall.

McClain, L. 1983. Behavior during examinations: A comparison of "A," "C," and "F" students. *Teaching of Psychology* 10: 69–71.

McClure, J., B. Sonak, and H. Suen. 1999. Concept map assessment of classroom learning: Reliability, validity, and logistical practicality. *Journal of Research in Science Teaching* 36: 475–492.

McDaniel, M., J. Anderson, M. Derbish, and N. Morisette. 2007. Testing the testing effect in the classroom. *European Journal of Cognitive Psychology* 19: 494–513.

McDaniel, M., A. Friedman, and L. Bourne. 1978. Remembering the levels of information in words. *Memory and Cognition* 6: 156–164.

McGuire, C. 1963. Research in the process approach to the construction and to analysis of medical examinations. *National Council on Measurement in Education Yearbook* 20: 7–16.

Michael, J. 2006. Where's the evidence that active learning works? *Advances in Physiology Education* 30: 159–167.

Michaelsen, L., A. Knight, and L. Fink. 2002. *Team-Based Learning: A Transformative Use of Small Groups.* Westport, CT: Praeger.

Michaelsen, L., A. Knight, and L. Fink. 2004. *Team-Based Learning: A Transformative Use of Small Groups in College Teaching.* Sterling, VA: Stylus Publishing.

Midkiff, K., T. Litzinger, and D. Evans. 2001. Development of engineering thermodynamics concept inventory instruments. *Proceedings of the Frontiers in Education Conference, Reno, NV.*

Momsen, J., T. Long, S. Wyse, and D. Ebert-May. 2010. Just the facts? Introductory undergraduate biology courses focus on low-level cognitive skills. *CBE—Life Sciences Education* 9: 435–440.

Montepare, J. 2005. A self-correcting approach to multiple-choice tests. *APS Observer,* 18.

Moog, R., F. Creegan, D. Hanson, J. Spencer, A. Straumanis, D. Bunce, and T. Wolfskill. 2009. POGIL: Process-Oriented Guided-Inquiry Learning. In *Chemists' Guide to Effective Teaching: Volume II.* N. Pienta, M. Cooper, and T. Greenbowe, eds. Upper Saddle River, NJ: Prentice Hall.

Moog, R. and J. Spencer. 2008. *Process-Oriented Guided Inquiry Learning: ACS Symposium Series 994.* Washington, DC: American Chemical Society.

Morgan, M., R. Clarke, M. Weidmann, J. Laidlaw, and A. Law. 2007. How assessment drives learning in neurosurgical higher training. *Journal of Clinical Neuroscience* 14: 349–354.

Morgan, C., J. Lilley, and N. Boreham. 1988. Learning from lectures: The effect of varying the detail in lecture handouts on note-taking and recall. *Applied Cognitive Psychology* 2: 115–122.

Morris, C., J. Bransford, and J. Franks. 1977. Levels of processing versus transfer appropriate processing. *Journal of Verbal Learning and Verbal Behavior* 16: 519–533.

Mulford, D. and W. Robinson. 2002. An inventory for alternate conceptions among first-semester general chemistry students. *Journal of Chemical Education* 79: 739–744.

Nakagawa, S. and I. Cuthill. 2007. Effect size, confidence interval, and statistical significance: A practical guide for biologists. *Biological Reviews Cambridge Philosophical Society* 82:591–605

National Research Council. 2003. *Bio2010: Transforming Undergraduate Education for Future Research Biologists.* Washington, DC: National Academies Press.

National Research Council (NRC). 2012a. *A Framework for K–12 Science Education: Practices, Crosscutting Concepts, and Core Ideas.* Washington, DC: National Academies Press.

National Research Council (NRC). 2012b. *Discipline-Based Education Research: Understanding and Improving Learning in Undergraduate Science and Engineering.* Washington, DC: National Academies Press.

Novak, J. 2005. Results and implications of a 12-year longitudinal study of science concept learning. *Research in Science Education* 35: 23–40.

Novak, J. and A. Cañas. 2006. The theory underlying concept maps and how to construct and use them. Florida Institute for Human and Machine Cognition. Retrieved from http://cmap.ihmc.us/Publications/ResearchPapers/Theory UnderlyingConceptMaps.pdf

Novak, G., A. Gavrin, W. Christian, and E. Patterson. 1999. *Just-in-Time Teaching: Blending Active Learning with Web Technology.* Upper Saddle River, NJ: Prentice Hall.

Novak, J. and D. Gowin. 1984. *Learning How to Learn.* Cambridge, MA: Cambridge University Press.

O'Donnell, A., D. Dansereau, and R. Hall. 2002. Knowledge maps as scaffolds for cognitive processing. *Educational Psychology Review* 14: 71–86.

Osborne, J. 2010. Arguing to learn in science: The role of collaborative, critical discourse. *Science* 328: 463–466.

Otter, S. 1992. *Learning Outcomes in Higher Education.* London: UDACE.

Palmer, E. and P. Devitt. 2007. Assessment of higher-order cognitive skills in undergraduate education: Modified essay or multiple-choice questions? *BMC Medical Education* 7: 49.

Palomba, C. and T. Banta. 1999. *Assessment Essentials: Planning, Implementing, and Improving Assessment in Higher Education.* San Francisco: Jossey-Bass.

Partnership for 21st Century Skills. 2009. Washington DC: Association for Supervision and Curriculum Development.

Pelaez, N. 2002. Problem-based writing with peer review improves academic performance in physiology. *Advances in Physiology Education* 26: 174–184.

Pellegrino, J., N. Chudowsky, and R. Glaser. 2001. *Knowing What Students Know: The Science and Design of Educational Assessment.* Washington, DC: National Academies Press.

Popham, J., E. Eisner, H. Sullivan and L. Tyler, eds. 1969. *Instructional Objectives: AERA Monograph Series on Curriculum Evaluation.* Chicago: Rand McNally and Company.

Preszler, R., A. Dawe, C. Shuster, and M. Shuster. 2007. Assessment of the effects of student response systems on student learning and attitudes over a broad range of biology courses. *CBE—Life Sciences Education* 6: 29–41.

Pungente, M. and R. Badger. 2003. Teaching introductory organic chemistry: Blooming beyond a simple taxonomy. *Journal of Chemical Education* 80: 779–784.

Race, P. 1996. The art of assessing 2. *New Academic* 5: 3–6.

Ramaprasad, A. 1983. On the definition of feedback. *Behavioral Science* 28: 4–13.

Ramsden, P. 1992. *Learning to Teach in Higher Education.* New York: Routledge.

Rasch, G. 1960. *Probabilistic Models for Some Intelligence and Attainment Tests.* Copenhagen, Denmark: Danish Institute for Educational Research.

Reeves, D. 2007. *Ahead of the Curve: The Power of Assessment to Transform Teaching and Learning.* Bloomington, IN: Solution Tree.

Robbins, R. 2009. Evaluation and assessment of career advising. In *The Handbook of Career Advising.* K. Hughey, D. Nelson, J. Damminger, and B. McCalla-Wriggins, eds. San Francisco: Jossey-Bass.

Robertson, R. 2001. Calibrated peer review. *American Biology Teacher* 63: 474–480.

Rodriguez, M. 2005. Three options are optimal for multiple-choice items: A meta-analysis of 80 years of research. *Educational Measurement: Issues and Practice* 24: 3–13.

Roediger, H. and J. Karpicke. 2006a. Test-enhanced learning: Taking memory tests improves long-term retention. *Psychological Science* 17: 249–255.

Roediger, H. and J. Karpicke. 2006b. The power of testing memory: Basic research and implications for educational practice. *Perspectives on Psychological Science* 1: 181–210.

Roediger, H. and E. Marsh. 2005. The positive and negative consequences of multiple-choice testing. *Journal of Experimental Psychology: Learning, Memory, and Cognition* 31: 1155–1159.

Roediger, H. and K. McDermott. 1993. Encoding specificity in perceptual priming. *Proceedings of the 9th Annual Meeting of the International Society for Psychophysics.* Madrid, Spain.

Roscoe, R. and M. Chi. 2004. The influence of the tutee in learning by peer tutoring. *Proceedings of the 26th Annual Meeting of the Cognitive Science Society.* London: Psychology Press.

Rowntree, D. 1982. *Educational Technology in Curriculum Development.* London: Harper and Row.

Sadler, D. 1989. Formative assessment and the design of instructional systems. *Instructional Science* 18: 119–144.

Schneps, M., P. Sadler, S. Woll, and L. Crouse. 1989. *A Private Universe*. S. Burlington, VT: Annenberg Media.

Schneps, M., P. Sadler, S. Woll and L. Crouse. 1997. *Minds of Our Own*. S. Burlington, VT: Annenberg Media.

Schroeder, J. and T. Greenbowe. 2008. Implementing POGIL in the lecture and the science writing heuristic in the laboratory—student perceptions and performance in undergraduate organic chemistry. *Chemistry Education Research and Practice* 9: 149–156.

Schulman, L. 2002. Making differences: A table of learning. *Change* 34: 36–44.

Scouller, K. 1998. The influence of assessment method on students' learning approaches: Multiple-choice question examination versus assignment essay. *Higher Education* 35: 453–472.

Shebilske, W., B. Goettl, K. Corrington, and R. Day. 1999. Interlesson spacing and task-related processing during complex skill acquisition. *Journal of Experimental Psychology: Applied* 5: 413–437.

Shepard, L. 2010. Next-generation assessments. *Science* 330: 890.

Shi, J., W. Wood, J. Martin, N. Guild, Q. Vicens, and J. Knight. 2010. A diagnostic assessment for introductory molecular and cell biology. *CBE—Life Sciences Education* 9: 453–461.

Shindler, J. 2004. Greater than the sum of the parts? Examining the soundness of collaborative exams in teacher education courses. *Innovations in Higher Education* 28: 273–283.

Slater, S., T. Slater, and J. Bailey. 2010. *Discipline-Based Education Research: A Scientist's Guide*. New York: W.H. Freeman.

Smith, S. and A. Glenberg. 1978. Environmental context and human memory. *Memory and Cognition* 6: 342–353.

Smith, M., C. Hinckley, and G. Volk. 1991. Cooperative learning in the undergraduate laboratory. *Journal of Chemical Education* 68: 413–415.

Smith, B. and J. MacGregor. 1992. What is collaborative learning? In *Collaborative Learning: A Sourcebook for Higher Education*. A. Goodsell, M. Mahler, V. Tinto, B. L. Smith, and J. MacGregor, eds. National Center on Postsecondary Teaching, Learning, and Assessment: Syracuse University.

Smith, A., R. Stewart, P. Shields, J. Hayes-Klosteridis, P. Robinson, and R. Yuan. 2005. Introductory biology courses: A framework to support active learning in large enrollment introductory science courses. *CBE—Life Sciences Education* 4: 143–156.

Smith, M., W. Wood, and J. Knight. 2008. The Genetics Concept Assessment: A new concept inventory for gauging student understanding of genetics. *CBE—Life Sciences Education* 7: 422–430.

Smith, M., W. Wood, W. Adams, C. Wieman, J. Knight, N. Guild, and T. Su. 2009. Why peer discussion improves student performance on in-class concept questions. *Science* 323: 122–124.

Springer, L., M. Stanne, and S. Donovan. 1999. Effects of small-group learning on undergraduates in science, mathematics, engineering, and technology: A meta-analysis. *Review of Educational Research* 69: 21–51.

Stearns, S. 1996. Collaborative exams as learning tools. *College Teaching* 44: 111–112.

Steif, P., A. Dollar, and J. Dantzler. 2005. Results from a statics concept inventory and their relationship to other measures of performance in statics. *Proceedings of the Frontiers in Education Conference, Indianapolis, IN.*

Stiggins, R., J. Arter, J. Chappuis, and S. Chappuis. 2007. *Classroom Assessment for Student Learning: Doing It Right—Using It Well.* Upper Saddle River, NJ: Pearson Education, Inc.

Streveler, R., R. Miller, A. Santiago-Roman, M. Nelson, M. Geist, and B. Olds. 2011. Rigorous methodology for concept inventory development: Using the "assessment triangle" to develop and test the thermal and transport science concept inventory (TTCI). *International Journal of Engineering Education* 27: 968–984.

Tanner, K. 2009. Talking to learn: Why biology students should be talking in classrooms and how to make it happen. *CBE—Life Sciences Education* 8: 89–94.

Tanner, K. 2012. Promoting student metacognition. *CBE—Life Sciences Education* 11: 113–120.

Taylor, K. and D. Rohrer. 2010. The effects of interleaved practice. *Applied Cognitive Psychology* 24: 837–848.

Thomas, A. and M. McDaniel. 2007. Metacomprehension for educationally relevant materials: Dramatic effects of encoding–retrieval interactions. *Psychonomic Bulletin and Review* 14: 212–218.

Thompson, M. and D. Wiliam. 2007. *Tight But Loose: Conceptual Framework for Scaling Up School Reforms.* Washington, DC: American Educational Research Association.

Thornton, R. and D. Sokoloff. 1998. Assessing student learning of Newton's laws: The Force and Motion Conceptual Evaluation and the evaluation of active learning laboratory and lecture curricula. *American Journal of Physics* 66: 338–352.

Timmerman, B. and D. Strickland. 2009. Faculty should consider peer review as a means of improving students' scientific reasoning skills. *Journal of the South Carolina Academy of Science* 7: 1–7.

Timmerman, B., D. Strickland, R. Johnson, and J. Payne. 2010. Development of a "universal" rubric for assessing undergraduates' scientific reasoning skills using scientific writing. *Assessment and Evaluation in Higher Education* 36: 509–547.

Treagust, D. 1988. Development and use of diagnostic tests to evaluate students' misconceptions in science. *International Journal of Science Education* 10: 159–169.

Trochim, W. 2013. *The Research Methods Knowledge Base.* 2nd ed. Retrieved from http://www.socialresearchmethods.net/kb/

Tyler, R. 1969. *Basic Principles of Curriculum and Instruction.* Chicago: University of Chicago Press.

Tyler, S., P. Hertel, M. McCallum, and H. Ellis. 1979. Cognitive effort and memory. *Journal of Experimental Psychology: Human Learning and Memory* 5: 607–617.

Udovic, D., D. Morris, A. Dickman, J. Postlethwait, and P. Wetherwax. 2002. Workshop biology: Demonstrating the effectiveness of active learning in an introductory biology course. *BioScience* 52: 272–281.

Vygotsky, L. 1981. The instrumental method in psychology. In *The Concept of Activity in Soviet Psychology*. J. Wertsch, ed. Armonk, NY: Sharpe.

Vygotsky, L. and M. Cole. 1978. *Mind in Society: The Development of Higher Psychological Processes*. Cambridge, MA: Harvard University Press.

Wage, K., J. Buck, C. Wright, and T. Welch. 2005. The signals and systems concept inventory. *IEEE Transactions on Education* 48: 448–461.

Walker, J., S. Cotner, P. Baepler, and M. Decker. 2008. A delicate balance: Integrating active learning into a large lecture course. *CBE—Life Sciences Education* 7: 361–367.

Walvoord, B. and V. Anderson. 1998. *Effective Grading: A Tool for Learning and Assessment*. 1st ed. San Francisco: Jossey-Bass.

Waterman, M. and E. Stanley. 1998. "Investigative cases and case-based learning in biology." A text module in the BioQUEST Library VI (on CD-ROM). New York: Academic Press.

Webb, N. 1989. Peer interaction and learning in small groups. *International Journal of Educational Research* 13: 21–39.

Wente, M. 2009. "We pretend to teach 'em, they pretend to learn." *The Globe and Mail*, April, 18.

Westbrook, J., A. Woods, M. Rob, W. Dunsmuir, and R. Day. 2010. Association of interruptions with an increased risk and severity of medication administration errors. *Archives of Internal Medicine* 170: 683–690.

Wiggins, G. and J. McTighe. 1998. *Understanding by Design*. Alexandria, VA: Association for Supervision and Curriculum Development.

Wiliam, D. and P. Black. 1996. Meanings and consequences: A basis for distinguishing formative and summative functions of assessment? *British Educational Research Journal* 22: 537–548.

Williams, A., N. Aguilar-Roca, M. Tsai, M. Wong, M. Beaupre, and D. O'Dowd. 2011. Assessment of learning gains associated with independent exam analysis in introductory biology. *CBE—Life Sciences Education* 10: 346–356.

Wilson, C., C. Anderson, M. Heidemann, J. Merrill, B. Merritt, G. Richmond, D. Sibley, and J. Parker. 2006. Assessing students' ability to trace matter in dynamic systems in cell biology. *CBE—Life Sciences Education* 5: 323–331.

Wilson, T. and L. Coyle. 1991. Improving multiple choice questioning: Preparing students for standardized tests. *Clearing House* 64: 422–424.

Wineburg, S. and J. Schneider. 2009. Inverting Bloom's Taxonomy. *Education Week* 29: 28–31.

Wolter, B., M. Lundeberg, H. Kang, and C. Herreid. 2011. Students' perceptions of using personal response systems ("clickers") with cases in science. *Journal of College Science Teaching* 40: 14–19.

Wood, W. 2004. Clickers: A teaching gimmick that works. *Developmental Cell* 7: 796–798.

Wood, W. 2009. Innovations in teaching undergraduate biology and why we need them. *Annual Review of Cell and Developmental Biology* 25: 93–112.

Woodford, K. and P. Bancroft. 2005. Multiple-choice questions not considered harmful. *Proceedings of the 7th Australian Conference on Computing Education.* Newcastle, Australia.

Woodin, T., V. Carter, and L. Fletcher. 2010. Vision and change in biology undergraduate education, a call for action: Initial responses. *CBE—Life Sciences Education* 9: 71–73.

Wormald, B., S. Schoeman, A. Somasunderam, and M. Penn. 2009. Assessment drives learning: An unavoidable truth? *Anatomical Sciences Education* 2: 199–204.

Yeo, S. and M. Zadnik. 2001. Introductory thermal concept evaluation: Assessing students' understanding. *The Physics Teacher* 39: 496–604.

Yuretich, R., S. Khan, R. Leckie, and J. Clement. 2001. Active-learning methods to improve student performance and scientific interest in a large introductory oceanography course. *Journal of Geoscience Education* 49: 111–119.

Zheng, A., J. Lawhorn, T. Lumley, and S. Freeman. 2008. Application of Bloom's taxonomy debunks the "MCAT myth." *Science* 319: 414–415.

Zoller, U. 1993. Are lecture and learning compatible? Maybe for LOCS; unlikely for HOCS. *Journal of Chemical Education* 70: 195–197.

Index

A

action research (AR). *See* AR (action research).

active learning, 61–62, 135

alignment. *See also* Bloom's Taxonomy.
    formative assessment with cognitive levels, 63
    of instructional materials, 9–10, 13
    of instructional practices, 22–25
    workshop module, 130–132

alignment table for a course, 135

alternative choice questions, 79, 91

alternative form reliability, 117, 135

ambiguous terms, 8–9

analytic rubrics, 49, 58

analyze, category of human cognition, 15

animations, 95, 103

apply, category of human cognition, 15

AR (action research). *See also* DBER (discipline-based education research).
    concept inventories, 111
    data collection instruments, selecting, 110–111
    definition, 135
    free-response questions, 111
    overview, 108
    qualitative *vs.* quantitative data, 109–110
    reliable instruments, 110–111
    rubrics, 111
    validated instruments, 110–111

AR (action research), research formats
    cross-over format, 113
    pre- and post-test format, 112, 113–116
    sequential course format, 112–113

"Ask, don't tell," 64

assessment. *See also* AR (action research); DBER (discipline-based education research).
    in backward design, 6
    critical functions, 4
    at the end of a learning session. *See* summative assessment.
    formative, 4–5
    forms of, 4–5
    in instructional design, 5–6. *See also* backward design.
    during a learning session. *See* formative assessment.
    preparing students for. *See* learning how to learn.
    summative, 4–5

assessment of learning gains
    across groups of students, 112–113
    all students in a class, 112
    individual students, 112

assessment workshop
    activities, 128–134
    alignment module, 130–132
    backward design module, 128–130
    Bloom's Taxonomy module, 131–132
    formative assessment module, 132–134
    intended learning outcomes module, 129–130
    materials needed, 125
    overview and goals, 126–127
    preparation, 125
    summative assessment module, 130–132

attention, effective learning, 94, 96

B

backward design
    aligning instructional materials, 9–10, 13
    ambiguous terms, 8–9
    assessment, 6
    basic steps, 6
    beyond the classroom, 11

backward design (*continued*)
    definition, 6, 135
    determining evidence of achievement, 8–9, 12
    evaluating the instruction, 6
    identifying learned outcomes, 7–8
    intended learning outcomes, 6
    planning learning experiences, 9
    planning the instruction, 6
    sample worksheet, 13
    transfer-appropriate processing, 9
    workshop module, 128–130
BLASt (Bloom's-based Learning Activities for Students tool), 98, 102
blocking time, 99–100
Blooming Biology Tool, 36–39
Bloom's Taxonomy. *See also* alignment.
    analyze, category of human cognition, 15
    apply, category of human cognition, 15
    Bloom T diagram, 16–17
    Bloom verbs, 15
    categories of human cognition, 15
    course alignment table, 22–25
    create, category of human cognition, 15
    definition, 135
    description, 15–17
    evaluate, category of human cognition, 15
    evaluating instructional alignment, 22–25, 76
    evolution of, 35
    examples, 37–39
    HOC (high-order cognition) skills, 15–17
    LOC (low-order cognition) skills, 15–17
    MCQs (multiple-choice questions), 78, 81
    in practice, 17–18
    protocol for applying, 19
    remember, category of human cognition, 15
    revised, 17–18, 35
    summary of educational taxonomies, 35
    testing single concepts at multiple levels, 20–21
    understand, category of human cognition, 15
    workshop module, 131–132
    writing questions with, 20–21
Bloom's Taxonomy, ranking questions
    description, 17–19

    sample questions, 26–34
Bloom T diagram, 16–17
Bloom verbs, 15
books and publications
    Classroom Assessment Techniques, 64
    Discipline-Based Education Research: A Scientist's Guide, 107
    Minds of Our Own, 63, 129
    A Private Universe, 7, 63, 129
    The Scarlet Letter, vii
    Scientific Teaching, viii, xi–xii, 128
    A Tiny World, 129
brainstorming, 66–67, 72, 135

C

Calibrated Peer Review (CPR™), 50, 59
case studies
    definition, 135
    small-scale formative assessment, 66, 68, 72
classical test theory, 135
Classroom Assessment Techniques, 64
clickers. *See* immediate response systems (clickers).
cognition. *See* human cognition.
cognitive levels
    aligning with formative assessment, 63
    of instructional material, evaluating. *See* Bloom's Taxonomy.
    of MCQs (multiple-choice questions), 76
    of questions, determining. *See* Bloom's Taxonomy.
Cohen's d, 118
collaborative learning, 65, 136
collaborative test-taking, 51–52. *See also* pyramid exams.
complex (K-type) questions, 79, 89
concept inventories
    AR (action research), 111
    definition, 136
concept inventory questions, 67–68
concept maps, 66, 69, 72, 95
constructivism, 62, 136
construct validity, 117, 136

content validity, 117, 136
context-dependent questions, 79, 86
cooperative learning, 65, 136
course alignment table, 22–25
CPR™ (Calibrated Peer Review), 50, 59
cramming, 99–100
create, category of human cognition, 15
cross-over research format, 113, 136

D

data collection instruments
    AR (action research), 110–111
    DBER (discipline-based education research),
        115–116
DBER (discipline-based education research). *See
    also* AR (action research).
    data collection instruments, 115–116
    definition, 136
    designing your study, 115
    ethical issues regarding human subjects, 119–
        120
    identifying good questions, 115
    overview, 114–115
    reliable instruments, 116
    validated instruments, 116
    valid instruments, 116
DBER (discipline-based education research), data
        analysis
    Cohen's d, 118
    effect size, 116–118
    IRT (item response theory), 119
deep approach to learning, 92–93
deliberate practice, 61–62
desirable difficulties, 100
Dewey, John, 62
diagnostic question cluster, 136
diagrams, 95
*Discipline-Based Education Research: A Scientist's
    Guide*, 107
discipline-based education research (DBER).
    See DBER (discipline-based education
    research).

discriminatory power of MCQs (multiple-choice
    questions), 78
distractors, MCQs (multiple-choice questions)
    definition, 77
    generating, 79
    quality of, 80
do-over tests, 53–54
DQCs (Diagnostic Question Clusters), 66, 68, 72
drawing exercises, 66, 69, 72
drawings, 95

E

educational taxonomies, summary of, 35. *See also
    specific taxonomies.*
effect size, 116–118, 136
elaboration, effective learning, 94, 95
"engaugement," 61–62, 136
essays, 45
ethical issues regarding human subjects, 119–120
evaluate, category of human cognition, 15
evaluating
    cognitive levels of instructional material. *See
        Bloom's Taxonomy.*
    cognitive levels of MCQs (multiple-choice
        questions), 76
    instruction, in backward design, 6
    instructional alignment, 22–25, 76
evidence of achievement, 8–9, 12
exam construction guide, 76, 136
Exam Wrapper technique, 98
expressive outcomes, xv

F

face validity, 117, 137
fatigue, effects on studying, 100
feedback
    in formative assessment, 63
    IF-AT (immediate feedback assessment tech-
        nique), 52–53, 59, 137
    in summative assessment, 50–51
fixed view of intellectual ability, 97

FLAG (field-tested learning assessment guide), 137

formative assessment. *See also* summative assessment.
  achieving intended learning outcomes, 63
  active learning, 61–62
  aligning with cognitive levels, 63
  "Ask, don't tell.," 64
  definition, 137
  deliberate practice, 61–62
  description, 4–5
  "engaugement," 61–62
  in practice, 64–65
  students' prior knowledge, 62–63
  timely feedback, 63
  vs. summative assessment, 44
  workshop module, 132–134

formative assessment, large scale
  MBL (model-based learning), 70, 73
  PBL (problem-based learning), 70, 73
  PLTL (Peer-Led Team Learning), 70, 73
  POGIL (process-oriented guided inquiry learning), 70, 73
  summary of types, 70. *See also specific types.*
  TBL (team-based learning), 70, 73

formative assessment, small scale
  brainstorming, 66–67, 72
  case studies, 66, 68, 72
  collaborative learning, 65
  concept inventory questions, 67–68
  concept maps, 66, 69, 72
  cooperative learning, 65
  description, 65–69
  DQCs (Diagnostic Question Clusters), 66, 68, 72
  drawing exercises, 66, 69, 72
  IF-AT (immediate feedback assessment technique), 66, 72
  immediate response systems (clickers), 66–68, 72
  JiTT (Just-in-Time-Teaching), 66–67, 72
  misconceptions, addressing, 68
  one-minute papers, 66, 69, 72

online quizzes, 69
post-class assessment, 69
quizzes, 66, 73
reading assessments, 66, 73
statement correction, 66, 73
strip sequences, 66, 69, 73
summary of types, 66, 73. *See also specific types.*
think-pair-share, 66, 73

free-response exams. *See also* free-response questions.
  description, 48
  pros and cons, 45, 48
  stereotype bias, 48
  subjective grading, 48

free-response questions. *See also* free-response exams.
  AR (action research), 111
  Bloom level, determining, 18
  converting to MCQs (multiple-choice questions), 81–84

G

general (holistic) rubrics, 49, 57
generation, effective learning, 94
goals. *See* learning goals.
goal-setting, 97
grading
  stereotype bias, 48
  subjective, 48
grading summative assessments
  CPR™ (Calibrated Peer Review), 50, 59
  peer grading, 50
  rubrics, 48–50
graphs, 95
grounded theory, 137

H

Handelsman, Jo, viii, xi
Hawthorne, Nathaniel, vii
highlighting missed material, 54
HOC (high-order cognition) skills, 15–17

holistic (general) rubrics, 49, 57
human cognition, categories of, 15. *See also*
        Bloom's Taxonomy.
human subjects, ethical issues, 119–120

I

IF-AT (immediate feedback assessment technique)
    definition, 137
    description, 52–53
    formative assessment, 66, 72
    online resources, 59
illusion of knowing, 92
immediate response systems (clickers), 66–68, 72,
        136
inquiry-based learning, 137
instruction, planning in backward design, 6
instructional alignment, evaluating, 22–25
instructional material, evaluating cognitive levels
        of. *See* Bloom's Taxonomy.
intellectual ability, fixed *vs.* fluid view of, 97
intended learning outcomes
    in backward design, 6
    definition, xvi, 138
    examples, 12
    formative assessment, 63
    workshop module, 129–130
interleaving
    definition, 137
    vs. blocking time, 100
internal consistency, 117, 137
interpretation, effective learning, 94, 95
interpretive research, 137
inter-rater reliability, 117, 137
intra-rater reliability, 117, 137
IRT (item response theory), 119
isomorphic items, 137
item analysis, 47
item difficulty, 117, 137
item discrimination, 117, 138
item response theory, 138
items, 117, 137

J

JiTT (Just-in-Time-Teaching), 66–67, 72, 138

K

keys, MCQs (multiple-choice questions), 77
knowledge maps, 95

L

Lanier, Judy, 130–131
large-scale formative assessment. *See* formative
        assessment, large scale.
learned outcomes, identifying, 7–8
learning experiences, planning in backward
        design, 9
learning gains, 138
learning goals, xv, 138
learning how to learn
    animations, 95, 103
    attention, 96
    BLASt (Bloom's-based Learning Activities for
        Students tool), 98, 102
    concept maps, 95
    deep approach, 92–93
    diagrams, 95
    drawings, 95
    effective learning components, 94
    effective learning strategies, 93
    Exam Wrapper technique, 98
    fixed view of intellectual ability, 97
    fluid view of intellectual ability, 97
    graphs, 95
    illusion of knowing, 92
    knowledge maps, 95
    maps, 95
    metacognition, 97–98
    mnemonics, 95
    multitasking, 96
    outlines, 95
    simulations, 96
    strategies for mastering course material, 93–94
    student planning and goal-setting, 97

learning how to learn (*continued*)
    summary sheets, 95
    superficial approach, 92–93
    visual representations of information, 95
learning how to learn, effective study sessions
    blocking time, 99–100
    changing study locations, 99–100
    cramming, 99–100
    desirable difficulties, 100
    interleaving *vs.* blocking time, 100
    massing time, 99–100
    overview, 98
    spacing across time and place, 99–100
    studying while fatigued, 100
learning objectives, xv–xvi, 138
learning outcomes, xv
LOC (low-order cognition) skills, 15–17

M

maps, 95
massing time, 99–100
matching questions, 79, 89
MBL (model-based learning), 138
MCQs (multiple-choice questions). *See also* multiple-choice exams.
    alternative choice, 79, 91
    Bloom level, determining, 18
    Bloom levels, 78, 81
    common formats, 79. *See also specific formats.*
    complex (K-type), 79, 89
    components of, 77
    context-dependent, 79, 86
    converting from free-response questions, 81–84
    correct options, 77
    discriminatory power, 78
    evaluating cognitive levels, 76
    examples, 86–91
    incorrect options, 77
    keys, 77
    matching, 79, 89
    multiple true/false, 79, 90

    standard, 79
    stems, 77
    structure and format, 77–78
    true/false, 79, 89
    two-tiered, 79, 87–88
    writing, 78–84
MCQs (multiple-choice questions), distractors
    definition, 77
    generating, 79
    quality of, 80
metacognition
    definition, 138
    effective learning, 94, 97–98
Miller, Sarah, viii, xi
Minds of Our Own, 63, 129
misconceptions, addressing, 68
mnemonics, 95
model-based learning (MBL), 138
The Montillation of Traxoline, 130–131
multiple-choice exams. *See also* MCQs (multiple-choice questions).
    analyzing questions, 47
    description, 47
    Exam Construction Guide, 76
    item analysis, 47
    negative testing effect, 47
    pros and cons, 45, 47, 75
    testing effect, 47
multiple-choice questions (MCQs). *See* MCQs (multiple-choice questions).
multiple true/false questions, 79, 90
multitasking, 96

N

negative testing effect, 47

O

objectives, xiv
one-minute papers
    definition, 138
    small-scale formative assessment, 66, 69, 72

online quizzes, 69

open-book exams, pros and cons, 46

oral exams, pros and cons, 45

outcomes
    expressive, xv
    learned, identifying, 7–8
    learning, xv

outcomes, intended learning
    in backward design, 6
    definition, xvi, 138
    examples, 12
    formative assessment, 63
    workshop module, 129–130

outlines, 95

P

PBL (problem-based learning), 138

peer grading, 50

peer-led team learning, 138

Pfund, Christine, viii, xi

PhET (Physics Education) group, 96

planning
    instruction in backward design, 6
    learning experiences in backward design, 9
    student planning and goal-setting, 97

POGIL (process-oriented guided inquiry learning), 138

Point-Recapture method, 55

portfolios, pros and cons, 46

post-test format. *See* pre- and post-test format.

practical exams, pros and cons, 45

pre- and post-test format, 112, 113–116, 138

preparing students for assessment. *See* learning how to learn.

presentations, pros and cons, 46

A Private Universe, 7, 63, 129

problem-based learning (PBL), 138

process-oriented guided inquiry learning (POGIL), 138

publications. *See* books and publications.

pyramid exams, 52. *See also* collaborative test-taking.

Q

qualitative *vs.* quantitative data, 109–110

questions. *See also specific question types.*
    analyzing, 47
    cognitive level, determining. *See* Bloom's Taxonomy.
    good, identifying, 115
    ranking with Bloom's Taxonomy, 17–19, 26–34
    writing with Bloom's Taxonomy, 20–21

quizzes
    formative assessment, 66, 73
    online, 69

R

reading assessments, 66, 73, 139

reflection in the learning process, 50–51

reliability
    alternative form, 117, 135
    definition, 139
    inter-rater, 117, 137
    intra-rater, 117, 137
    rubrics, 122–123
    test-re-test, 117, 139

reliable instruments
    AR (action research), 110–111
    DBER (discipline-based education research), 116

remember, category of human cognition, 15

research formats. *See also* AR (action research); DBER (discipline-based education research).
    cross-over format, 113
    pre- and post-test format, 112, 113–116
    sequential course format, 112–113

Research Methods Knowledge Base, 113

retrieval practice, effective learning, 94

revised Bloom's Taxonomy, 17–18, 35

rubrics
    analytic, 49, 58
    AR (action research), 111
    creating, 49, 122–123

rubrics (*continued*)
  definition, 48
  general (holistic), 49, 57
  grading summative assessments, 48–50
  resources for, 59
  types of, 49–50. *See also specific types.*
  validity and reliability, 122–123

S

The Scarlet Letter, vii
Scholarship of Teaching and Learning (SoTL), 107
Science, Technology, Engineering and Mathematics (STEM), 139
scientific teaching, 139
Scientific Teaching, viii, xi–xii, 128
scoring. *See* grading.
self-corrected exams, 53
sequential course format, 112–113, 139
simulations, 96
small-scale formative assessment. *See* formative assessment, small scale.
SoTL (Scholarship of Teaching and Learning), 107
spacing effect, 139
spacing study sessions across time and place, 99–100
statement correction, 66, 73
stems, MCQs (multiple-choice questions), 77
STEM (Science, Technology, Engineering and Mathematics), 139
stereotype bias, 48
strip sequences
  definition, 139
  small-scale formative assessment, 66, 69, 73
student learning and summative assessment, 44
students
  as empty vessels, 62
  planning and goal-setting, 97
  preparing for assessment. *See* learning how to learn.
  prior knowledge, formative assessment, 62–63

studying, effective techniques
  blocking time, 99–100
  changing study locations, 99–100
  cramming, 99–100
  desirable difficulties, 100
  interleaving *vs.* blocking time, 100
  massing time, 99–100
  overview, 98
  spacing across time and place, 99–100
  studying while fatigued, 100
subjective grading, 48
summary sheets, 95
summative assessment. *See also* formative assessment.
  definition, 139
  description, 4–5
  do-over tests, 53–54
  feedback in the learning process, 50–51
  key elements, 56
  as a learning tool, 50–51
  in practice, 44–46
  prior to returning the test, 53–54
  reflection in the learning process, 50–51
  and student learning, 44
  vs. formative assessment, 44
  workshop module, 130–132
summative assessment, after returning the test
  highlighting missed material, 54
  Point-Recapture method, 55
  test analysis, 55
summative assessment, during the test
  collaborative test-taking, 51–52. *See also* pyramid exams.
  IF-AT (immediate feedback assessment technique), 52–53, 59
  pyramid exams, 52. *See also* collaborative test-taking.
  self-corrected exams, 53
summative assessment, forms of. *See also specific forms.*
  essays, 45
  free-response exams, 45
  multiple-choice exams, 45

open-book exams, 46
oral exams, 45
portfolios, 46
practical exams, 45
presentations, 46
summary of, 45–46
written reports, 46
summative assessment, grading
    CPR™ (Calibrated Peer Review), 50, 59
    peer grading, 50
    rubrics, 48–50
superficial approach to learning, 92–93

T

taxonomies (educational), summary of, 35. *See also specific taxonomies.*
team-based learning, 139
test analysis, 55
testing effects, 47, 94, 139
test-re-test reliability, 117, 139
A Tiny World, 129
T-P-S (think-pair-share)
    definition, 139
    small-scale formative assessment, 66, 73
transfer-appropriate processing, 9
true/false questions, 79, 89
two-tiered questions, 79, 87–88

U

understand, category of human cognition, 15

V

validated instruments
    alternative form reliability, 117
    AR (action research), 110–111
    construct validity, 117
    content validity, 117
    in DBER, 116
    definition, 139
    examples, 121–122
    face validity, 117
    internal consistency, 117
    inter-rater reliability, 117
    intra-rater reliability, 117
    item difficulty, 117
    item discrimination, 117
    items, 117
    reliability, 117
    test-re-test reliability, 117
    validity, 117
validity
    construct, 117
    content, 117
    in DBER, 116
    definition, 139
    face, 117
    rubrics, 122–123
visual representations of information, 95

W

worksheets, backward design, 13
workshop. *See* assessment workshop.
written reports, pros and cons, 46

About the Authors

Clarissa Dirks is an Associate Professor of Biology at The Evergreen State College in Olympia, Washington. She earned her PhD in Molecular and Cellular Biology at the University of Washington, conducting research in virology at the Fred Hutchinson Cancer Research Center. She currently investigates the evolution of viruses and host viral-inhibitory proteins, as well as the distribution and biodiversity of Tardigrada species. As a Biology Education Researcher, she has implemented programs to improve retention of underrepresented students in first year science courses, conducted studies to better understand how students acquire and master science process and reasoning skills, and is developing assessment instruments to measure undergraduates' science process skill acquisition. She has received two Tom Rye Harvill Awards for the Integration of Art and Science, has been named a National Academies Education Fellow and Mentor in the Life Sciences, and is the recipient of two Biology Leadership Education grants. She works to provide professional development opportunities for faculty and post-doctoral scholars by serving on the National Academies Summer Institute for Undergraduate Science Education Committee, leading a Pacific Northwest Regional Summer Institute, and mentoring post-doctoral fellows as a regional field station leader for the Faculty Institute for Reforming Science Teaching. As a member of the National Research Council's Board on Life Sciences committee on Developing a Framework for an International Faculty Development Project on Education about Research in the Life Sciences with Dual Use Potential, she trains faculty in best practices for teaching responsible conduct of research in their countries. She is a member of the Editorial Board of the journal *CBE—Life Science Education* and a co-founder of the Society for the Advancement of Biology Education Research (SABER).

Dr. Mary Pat Wenderoth is a Principal Lecturer in the Department of Biology at the University of Washington and teaches upper division animal physiology courses. She is a member of the University of Washington Biology Education Research Group, a group of twenty to thirty faculty, post-docs, graduate students, and undergraduate students who meet weekly to discuss the impact of innovative active learning practices on student learning. Dr. Wenderoth won the University of Washington Distinguished Teaching Award in 2001 and is a member of the University of Washington Teaching Academy. Dr. Wenderoth has been involved with the faculty development efforts of the National Academies Summer Institute on Undergraduate Teaching (now known as the National Academies Scientific Teaching Alliance, NASTA) since 2006 and continues to be involved with the new regional summer institutes that began in 2011. In 2010, Dr. Wenderoth co-founded the Society for the Advancement of Biology Education Research (SABER). SABER is a national network of faculty, post-docs, and graduate students who are conducting hypothesis-driven research in an effort to create a body of evidence-based teaching practices for undergraduate biology courses.

Michelle Withers is an Associate Professor of Biology at West Virginia University. Her research focuses on improving undergraduate science education, particularly evaluating the efficacy of different teaching methods in enhancing student learning. Another major focus of her program is training faculty and future faculty in scientific teaching. She runs the National Academies Summer Institute at West Virginia University, a regional offshoot of the National Academies Summer Institute on Undergraduate Biology Education (NASI). She serves as the Director for the National Academies Scientific Teaching Alliance (NASTA), and on the executive board of the Biology Director's Consortium (BDC), and is a founding member of the Society for the Advancement of Biology Education Research (SABER). She graduated with a BS in Public Health from the University of North Carolina at Chapel Hill and received her PhD in Neuroscience from the University of Arizona, Tucson.

Special thanks to Nancy Pelaez from Purdue University for reviewing and providing input on early drafts of this book.

The Scientific Teaching Book Series

The Scientific Teaching Book Series is a collection of practical guides, intended for all science, technology, engineering, and mathematics (STEM) faculty who teach undergraduate and graduate students in these disciplines. The purpose of these books is to help faculty become more successful in all aspects of teaching and learning science, including classroom instruction, mentoring students, and professional development. Authored by well-known science educators, the Series provides concise descriptions of best practices and how to implement them in the classroom, the laboratory, or the department. For readers interested in the research results on which these best practices are based, the books also provide a gateway to the key educational literature.

For ongoing information and discussions regarding this and other Scientific Teaching books, please visit: www.whfreeman.com/facultylounge/scientificteaching.

Co-editors for the Scientific Teaching Series:
- Sarah Miller, Madison Teaching and Learning Excellence, University of Wisconsin–Madison, Madison, Wisconsin
- William B. Wood, Science Education Initiative and Department of Molecular, Cellular, and Developmental Biology, University of Colorado, Boulder, Colorado

For further information about the series, please contact:

Susan Winslow, Publisher
Macmillan Higher Education
41 Madison Avenue, New York, NY 10010
swinslow@whfreeman.com

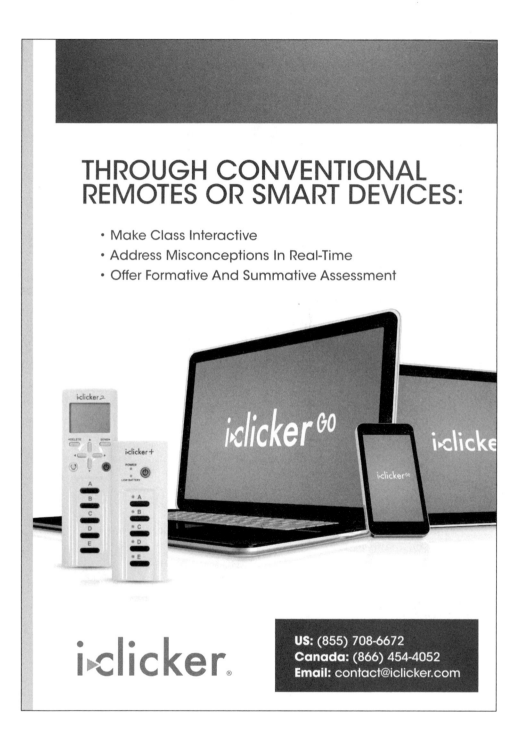